MANAGEMENT OF TECHNOLOGICAL CHANGE

MANAGEMENT OF TECHNOLOGICAL CHANGE

The Great Challenge of Management to the Future

by

ERNST G. FRANKEL

Department of Ocean Engineering,
Massachusetts Institute of Technology,
Cambridge, MA, U.S.A.

KLUWER ACADEMIC PUBLISHERS
DORDRECHT / BOSTON / LONDON

Library of Congress Cataloging in Publication Data

Frankel, Ernst G.
Management of technological change : the great challenge of management for the future / by Ernst G. Frankel.
p. cm.

1. Technological innovations--Management. I. Title.
HD45.F716 1990
658.5'14--dc20 90-4077

ISBN-13: 978-94-010-7389-9 e-ISBN-13: 978-94-009-1988-4
DOI: 10.1007/978-94-009-1988-4

Published by Kluwer Academic Publishers,
P.O. Box 17, 3300 AA Dordrecht, The Netherlands.

Kluwer Academic Publishers incorporates
the publishing programmes of
D. Reidel, Martinus Nijhoff, Dr W. Junk and MTP Press.

Sold and distributed in the U.S.A. and Canada
by Kluwer Academic Publishers,
101 Philip Drive, Norwell, MA 02061, U.S.A.

In all other countries, sold and distributed
by Kluwer Academic Publishers Group,
P.O. Box 322, 3300 AH Dordrecht, The Netherlands.

Printed on acid-free paper

Softcover reprint of the hardcover 1st edition 1990

TABLE OF CONTENTS

LIST OF FIGURES

Page

LIST OF TABLES

ACKNOWLEDGEMENTS

This book is the outcome of years of concern with the impact of ever more rapidly changing technology and the problems of managing these changes. The success of ventures is ever more dependent on timely and often courageous technological change decisions. I have learned from bitter experience that the use of traditional concepts of operational and economic life of technologies must be used with caution and that the real and impending impact of technological change is an ever present factor.

Many colleagues and friends contributed to my understanding of the management of technological change. Professor Y. S. Chang of Boston University was my mentor during much of this work and assisted me greatly in the understanding of the issues involved in the management of technological change. I am similarly indebted to many of my friends in the MIT Program on Management of Technology.

Most importantly, my thanks go to Ms. Sheila McNary who patiently and painstakingly compiled, drafted, and edited this work. Any errors or omissions obviously are my sole responsibility.

Ernst G. Frankel
Cambridge, Massachusetts USA

FOREWORD. The competitive position of the older industrial nations, like those in Western Europe and more recently the U.S., is deteriorating, notwithstanding increasing expenditures for research, development, and education, and investment in productive assets. A large percentage of inventions and discoveries made in these countries do not lead to new processes or products or only after a long period of time. In fact the majority of inventions or discoveries in these countries are never even subjected to an effective innovation or development process and, as a result, actual use. Those that are, quite often take many years before being brought into use because of the slow progress of the innovation and implementation steps, lack of or insufficient support, conflicting interest of the underwriters who often prefer to hide than develop a new technology, or the perceived lack of interest by the market. A major reason for such delays in most western industrialized countries is the fragmentation and lack of continuity in the steps leading from research and development to invention or discovery and from innovation to first possible use, and finally refinement, implementation, or diffusion of the new product or process. New technology therefore develops more slowly in many western countries, even though many new technologies were invented and often developed in these same countries. The new industrial nations of the Far East, on the other hand, appear to use much more streamlined continuous technology development and applications procedures which appear to provide opportunities for much more rapid technology development.

In this book technology is defined as physical or operational products or processes which require specific design or procedure. Technology fundamentally consists of knowledge of what to do and how to design and use a service, product, process, and procedure. It is at the heart of all forms of economic activity no matter how primitive or sophisticated. The initiation of any type of economic activity starts with the acquisition of knowledge from within or without, towards an economic objective. The use of technology constitutes the performance of an economic activity, and technological change constitutes a change in economic activity. Such a technological change may cause improvements in productivity which require a reallocation of resources and revaluation of outputs whose costs may have changed. Therefore a technological change may have negative effects such as the reduction of labor required for production of an output, while at the same time reducing its costs and thereby (if price follows costs) make it more accessible.

The management of technology today drives marketing, design, manufacturing, and ultimately a firm's performance. It should be properly integrated with people management to provide effective solution to the challenges and opportunities facing us today. Technology changes the working environment, the work content, work condition, and obviously output, which in turn affects the value of worker output. Technological change may result in fewer but much better jobs, as it often changes work content. It may cause generation of new business and job opportunities which may ultimately replace or generate more and better jobs than were lost by the initial technological change.

The most important task of management in industrial and service organizations today is to learn how to manage technological change and the resulting challenges in people management and marketing. This requires a new type of manager, one who is concerned with financial and market performance but recognizes that these are the effects of his ability to manage the causes of a firm's performance which today is solidly founded on the effective management of technological change.

THE ROLE OF TECHNOLOGICAL CHANGE

There are several determinants of technological change. Technological change affects processes and products. It is the advance of technology which may consist of new methods for producing existing products, new product designs to permit improved production of existing products, new products with important new characteristics, as well as new approaches to management, control, organization, and marketing which constitute or involve technological improvements or change.

Technological changes constitute advances in knowledge, not just introduction of new techniques. They may involve new scientific principles discovered through scientific research or otherwise, but technological change involves use and the improvement of process and/or product, not only discovery of new knowledge.

Changes in technology may be connected with scientific discoveries or technological innovations but they usually do not follow these in a simple, direct manner. In fact, scientific discoveries may lay dormant or are applied in completely unrelated fields before they are used to support a change in a particular technology.

Today, technological change is more often based on scientific discoveries than in the past when technological change was often the result of chance discovery, trial and error experimentation, or reasoning. The rate of scientific discovery and the speed by which such discoveries are brought into product and process use has also accelerated which imposes new demands on management. Technological change often offers introduction of completely new products which in turn may result in important technological changes in other products, in services, and in processes. An example is development of computers for data processing and their use in manufacturing and processing industries, and in the development of new computer-controlled power or processing plants.

The rate of development and use of technological change, including technology transfer, is affected by the economic or profit advantage over older products or processes, the risk involved in adopting the technology change, the amount of resources required for its development and the associated uncertainties in resource commitments, and finally the risk that the technology change will perform as expected.

Factors causing technological change are often difficult to determine. It is generally assumed that the introduction of technological change is influenced by profit potentials which in turn affect the rate of expenditure for factors needed to introduce the technological change. Similarly, there are supply factors which influence the cost of making particular kinds of technological change. While technological change in the pre-World War II period was largely the result of gradual evolution, and marginal change, it is now much more radical.

Investments for fixed assets in the western industrial countries are usually justified on the basis of short- to medium-term returns. Investments were often delayed until obsolescence of existing assets or demand for a new asset by market or competitive pressures. Long-term development, long-term financial return, or long-term market position have less frequently driven such investment in the west. In the Far East, on the other hand, the rate and not condition of technical obsolescence and projected long-term market changes are major forces influencing such investments.

In the U.S., the world's largest industrial economy, nearly 50% of all scientists and engineers work for the federal government which, together with the defense and space industry, accounts for over 75% of all scientists and engineers. The rest of U.S. industry, private research organizations, etc. account for under 20% of the total, with about 6% employed by academia. Yet the bulk of all inventions or discoveries which affect the economy of the U.S. are not made by the federal government, the defense or space industry or even academia, but by other, usually small, companies or research institutions. Even the large, mature, traditional industries, such as steel, textiles, automobiles, chemicals, etc., account for a disproportionate number of scientists and engineers to the number of inventions made or innovations introduced. It is by and large the small, often entrepreneurial companies (or individuals) who account for the vast majority of new inventions and ultimately innovations and technological developments in the western world.

It is quite different in the Orient where industrial firms, academia, government, and research laboratories cooperate much more closely. The problem western industrial nations face today is two-fold: the management of technological change and the management of people change. Managers in the west know how to manage money but usually have less competence in and concern with the management of technology or people, tasks that are commonly left to the responsibility of lower level managers. In the Orient, technological and people management are the principal responsibilities of top management, usually comprised of professional engineers or technologists, while financial management is delegated to a lower level. This approach permits greater flexibility in adapting training, incentives, promotion and organizational structures to changing technologies, and eliminates much of the disjointed, often uninformed and belated, personnel and technological decisions made in the west.

FORMS OF TECHNOLOGICAL CHANGE

Technological change can occur in a process, product, or service. When there is a technological change in the product, then the processes used in its production may require technological change as well. For example, introduction of all welded ships caused a major technological change in shipbuilding affecting steel fabrication and erection processes.

Technological changes may affect capital and labor inputs in different proportions and are therefore often defined as capital or labor-saving technological changes. This obviously applies mainly to technological changes in processes or services in which performance is mainly measured in terms of use of labor and capital inputs. Technological change in products though may provide a change in product performance which bears no relationship to the output or performance of other products in terms of inputs. This may be caused because the output or performance is radically different from that of any other product or because the product is designed to be used differently.

One of the factors inducing or encouraging technological change is the desire for productivity growth, usually measured in terms of growth of output as a function of inputs. The potential for productivity growth may also further technological change by technological diffusion or technology transfer. Where technological change affects the product or its performance, market factors, including competition, may provide the driving force. Productivity growth is more difficult to measure than the growth of productivity in manufacturing because the product is designed to perform a service, and service performance improvements are not directly related to the inputs used.

The application of a new technology and its use may also be affected by the often highly fluctuating prices of the inputs and outputs, as well as market conditions, which may discourage productivity growth and encourage continued use of often obsolete technology, particularly when the old technology is depreciated or has a low financial cost and other inputs must be utilized, however inefficiently.

Technological change is often justified on the basis of productivity growth or cost savings. Yet productivity which relates changes in output to changes in inputs is often found to be an insufficient measure in the determination of the effects of technological change, because productivity growth may be affected more by better methods, organization, and management than by technological change. These are often closely linked or interdependent.

We also have a problem in measuring the rate and the magnitude of technological change as indirect measures, such as the effects of technological change, are generally used. Since they equate the effects of technological change with whatever increase in output is unexplained by other factors, they do not isolate the effects of technological change alone. We also often experience difficulties in isolating and measuring inputs.

Civilization and technology are intimately interwoven. In fact, human development is largely a progression of technological discoveries or technological changes. The use of sticks and flintstones in the manufacture of tools and the lighting of fire were among man's early technological discoveries which later led to the construction of artificial shelters and the development of agricultural implements. In turn these allowed man the freedom of growing food where needed and not where available through nature.

Early technology usually lacked a scientific and engineering base and was largely the result of chance discovery or trial and error experimentation. As a result, its growth was very slow and incremental. About 3000 years ago, in some parts of the world, man learned to perform more systematic investigations by experimentation and later scientific analysis as evidenced by Chinese, and later Greek, scientific advances. The industrial revolution finally led to formal scientific methods of technological development supplemented by engineering approaches designed to reduce scientific results and discoveries to use or apply.

Within the last few hundred years, a very short time on a historical scale, the rate and nature of technological advances has changed radically with technological advances aimed less at known or perceived needs but at advancement of knowledge and capability. This would often generate completely new demands or needs which, in turn, encouraged further investigation and discovery. While pre-industrial technology usually lacked both an extensive scientific and engineering base, modern technology is characterized by rapid and often dramatic changes and growth. With its roots in scientific research and drive for new knowledge, it can apply sophisticated scientific methods and use powerful engineering tools for its development. Technological advances have become more frequent, and few technological developments reach their economic or physical life before obsolescence. Technological change has become an ever increasingly important phenomenon affecting all aspects of human life and endeavor, and can no longer be left to chance, but must be managed to assure more effective, more timely, and more responsible use of technology.

The major problems in managing technological change are the selection of new technology, evaluation of its expected performance, timing of the introduction of the new technology, and determination of the rate and scale of introduction of the new technology. Management of technological change is therefore a complex and dynamic decision problem, because most parameters affecting the decision are constantly changing. In this book, technological forecasting, technology assessment, choice, and management techniques are presented which, it is hoped, will make management of technological change, easier, and more effective.

Ernst G. Frankel
Cambridge, Massachusetts
June 1989

1. Introduction

Technological change has been recognized as the major contributor to economic growth and has become one of the most important challenges to policy makers and managers. Many excellent books and papers have been written on the subject. Most of these deal with the macro or micro economic impact of technological change or the technological change process from invention and discovery to innovation, development, and final maturity as well as ultimate obsolescence of technology.

This book is designed to present technological change as a decision process and explain the use of recently developed methods for the effective management of technological change. In particular, techniques for the effective choice among technological alternatives, timing of the introduction of new technology both in terms of its own status and that of the technology to be replaced if any, and the rate and method of introduction of new technology are presented.

Management of technology is a complex decision process which is affected by both internal and external factors. The purpose of this book is to instruct the reader in effective technology decision making which involves the evaluation of the status of technology in use if any, the problem to be solved or output to be obtained, determination of environmental and internal constraints, and the competitive environment or market conditions which affect the technology decisions. Next, the existing and potential availability of technology of use must be determined by technological forecasting and status determination in which the range and condition of development of potentially useful technology is identified.

Methods for computing the economic and financial impact of technological change are developed, including the determination of cost, productivity, and quality obtained in the use of a technology. Similarly, to permit effective evaluation of the capability of technology, sources and methods for the acquisition of technology are reviewed and modes of technological change discussed. This is important as the condition of a technology during the innovation process, when it is refined from an initial invention or discovery to something useful, is a major factor influencing both choice and timing of technological change or adoption of the new technology.

Another issue of concern is the impact of market competition which in turn is affected by competitor's productivity or costs, quality, and strategy as well as technological diffusion, a measure of penetration, or extent of use of a technology.

The use of decision theory, including deterministic and probabilistic cross impact models, is shown to provide an effective means for relating the various factors influencing the choice, timing, and rate of introduction or the management of technological change.

Similarly, hierarchial analytical decision process models are shown to provide an effective structure for the consideration of diverse objectives of various decision makers who may influence the management of technological choice, timing, and rate of introduction. This approach also permits incorporation of environmental, financial, or economic constraints and their effect on the management of technological change.

The main objective of this book is to develop a formal, reliable, and correct approach to the complex evaluation and forecasting of technology and formulation of the change decision process which considers and in fact uses all the factors which have been identified as influencing that process. To accomplish this objective, Chapter 1 presents a general review of the issues involved in the technological change process and in the management of technological change. In Chapter 2 the research of various investigators concerned with issues of technological change is described and evaluated. It is shown that recent research placed major emphasis on the macro economic impact of technological advance and the micro economic effect of technological change at the firm's level.

To show the economic effects of technological change, production economics using production function analysis and micro economic theory in general is introduced in Chapter 3. The concept of technical substitution and elasticity are used to compare the effect of alternative technologies.

Chapter 4 covers the broad area of technology performance, evaluation, and productivity measurement. The role of experience or learning curves and technical progress functions in the valuation of the effects of technological change are discussed, and the shortcomings and abuses of learning or experience curve analysis as productivity or cost trend measures are reviewed.

Models of endogenous and exogenous technological change are presented in Chapter 5, together with analysis of factor input and augmentation.

Chapter 6 covers the sources of new technology, methods used in technology acquisition, including technology transfer, licensing, purchasing, as well as ways of protecting technological advances by patents and other means. The effectiveness of modern patenting or other protective measures used is evaluated. Technology pricing and marketing and their role in technology diffusion are reviewed in Chapter 7. The impact of maturity on technology diffusion strategy are also discussed, and technology flow among industries is reviewed.

In Chapter 8 we review methods of technological forecasting, its organization, validity, and use. Process, service, and product technology innovation cycles and their use are discussed in Chapter 9, in which capacity expansion and cost reduction problems are used to introduce models and solution techniques for management decisionmaking such as optimal process or product choice over time.

Chapter 10 covers the theory and methods of decision theory, analytical hierarchical processes, and their use in the management of technological change. The application of utility theory to the solution of corporate management decision problems is also reviewed and the use of cross-impact technology forecasting in hierarchical decision process analysis of technological change is presented here.

The effects of technological change on the economy and society as well as the international flow of technology are discussed in Chapter 11, as is the comparative management of technological change. Finally, Chapter 12 is an attempt at forecasting the future role of the management of technological change in economic, social, and environmental growth, at a time when we are increasingly surrounded by technology and technology permeates all aspects of the life of individuals, communities, nations, and the world. The effectiveness of management of technological change now affects the economic growth of nations, but in the future will affect the quality of the world's environment and its ability to sustain a high quality of human life. In summary, the purpose of this book is to present methods and techniques for effective management of technological change.

Technology is as old as the human race, and is one of the developments which set humans apart from other beings. Within the last century or two, a very short time on a historical scale, not only the rate but the very nature of technological change has been modified.

Pre-industrial technology usually lacked both extensive scientific as well as a sound engineering base, and its growth was slow and incremental. It was largely developed by trial and error and was mainly concerned with the replacement of muscle by motive power. Modern technology, on the other hand, is characterized by more continuous and rapid change. It often uses scientific principles for discovery and invention and sound engineering for engineering of the innovation.

Today effective understanding and management of technological change is essential for progress and economic growth at the firm's as well as the national level. The objectives of this book are to describe the processes involved, discuss the economic impact of technological change at the micro and macro level, and to develop analytical methods for the effective management of technological change from invention to innovation and commercialization. It is hoped that the reader will find the subject exciting, the methods useful, and the approach used instructive.

1.1 TECHNOLOGICAL CHANGE

Technological change poses both threats and opportunities. The rate at which new inventions appear and technological innovation are developed has greatly accelerated in recent years. Technological change has become the dominant factor in productivity improvements. Similarly technological change is a major contributor to industrial development. Today it is among the most important issues facing industrial management, and for that matter government. Roger B. Smith[1], Chairman of G.M.C., emphasized that the "greatest need for a company today is the management of change and, in particular, technological change which requires effective technological planning and goal orientation". Many alternative strategies for the management of technological change have been proposed. The choice of management strategy for technological change depends, among others, on the impact of technological change on productivity, quality, capacity, use of resources consumed, and related issues. The rate of technological change has increased rapidly in most industries during recent years. The strategy used in introducing technological change and the effectiveness of management of the technological change process often influence the success or failure of technological transition and business success. Yet, we often find that the management of technological change is haphazard. Only a few firms have developed or use plans for dealing effectively with technological change. As a result, there are large intra- and inter-industry differences in the effectiveness of management and resulting impact of technological change.

There are many reasons for the lack of effective planning for and management of technological change. Among these is the real and perceived risk of technological change, including the assumed uncertainties involved in the diffusion and acquisition of technology, lack of adequate technology evaluation, insufficient control of the rate of technological change and introduction, uncertainties in timing, competitiveness, market share, client perception, client requirements, company technology level, and other factors. The more rapid the rate of technological change the more resources must usually be committed to the process of technological change and, as a result, the risk is usually assumed to be larger.

---

[1] L. Burns, "Sky", interview with Roger B. Smith, Chairman of G.M.C., Vol. 12, January 1984.

Measures used to determine the rate of technological change are usually indirect and only consider the effects of technological change, as shown by Mansfield [1] in his original work on the economics of technological change and its impact on industry.

A number of industry studies of the management of technological change; such as Kekala [2] who evaluated the steel industry and explained the process of technological change in terms of its impact on steel-making productivity; Enos [3] who studied the oil refining industry from an historic development perspective in which the relations among the inventors, developers, and users are explained; and, Hollander [4] who performed a formal microeconomic analysis of rayon manufacture at several plants of E. I. duPont deNemours, were early attempts at determining the effects of technological change on industrial advance.

Other researchers, such as Griffin [5], modelled the impact of technological change on electric power generation using microeconomic models in which exogenous factors were incorporated in the demand function. These and other researchers studied particular industries to determine the sources or causes of increased efficiency and the methods of introduction and resulting impacts of technological change at the plant level. Although some of these researchers considered the impact of technological change in processes, the analysis dealt exclusively with the impact on the plant or industry.

The institutional approach towards technological change in different countries is usually quite diverse. While some technological change is recognized as an important element in the framework of industrial development, its role often evolves quite differently. While some countries attempt to use research as a means for the achievement of technological advance, others, at least during formative years, import technology and adapt it to their requirements. The organization of and relations between labor and management are similarly quite diverse, as are the relations between industry and government, and towards social acceptance of change.

Another important difference among these countries is the diverse development of corporate culture and infrastructure. This not only affects the timing of technological change and the time lag before technology change adoption, but also affects management and organizational resistance to technological change. In turn, such resistance affects the cost of technology change and its wider diffusion. This is amplified by the lower cost of introduction of technological change which, in turn, enhances wider distribution of technological change. This, in turn, causes economies of scale of technological change and further reduces its costs.

The above are some of the factors which may influence the effectiveness of management of technological innovation and change. Other factors appear to be government support of research and development, government policy towards the import of technology, government aid and assistance to industry, and government/industry relations. Similarly management structure, ownership, and control of a firm may influence the management of technological change.

Emphasis is increasingly placed upon technological process change as a means for growth, increased productivity, lower unit cost, and better ability to cope with product changes. there is a concern that the acceleration of technological product and process change and the expanded commitment of resources for product and process innovation may increasingly result in obsolescence of newly-introduced technology before effective amortization of the investment used. This has a profound impact on the choice, timing, and rate of introduction of new technology.

The strategy that management uses today to cope with technological change affects industry more than ever before and sometimes poses a substantial threat if many of the product and process changes and related technological advances come from outside the main stream of the industry.

Technology can be defined as the knowledge, experience, knowhow, and physical equipment or facilities which can create certain products or produce new types of products and services.

There is continuous pressure to improve products, processes, and services. Technological change is therefore increasingly driven by market pressures, competition, national and personal pride, scientific prowess, and chance. Today technology affects business as well as individuals, firms, and nations.

The most important method used in the advance of technology is innovation. Innovation is defined as the development of an idea, invention, or discovery towards commercial exploitation, by improving its ability to serve market demands. Innovation is therefore a process whereby ideas, inventions, or discoveries are transformed into more usable and marketable products, processes, or services.

Innovation may be driven by the recognition of a new idea which stems from an opportunity perceived in the market place, or a new technological development which may offer new uses, markets, improved performance, and greater competitiveness. Innovation as a process is driven by basic factors such as

1. a need or potential demand,
2. technical or operational means for satisfying it,
3. financial, technical, and operational means for undertaking the innovation,
4. timeliness of innovation, and
5. access and rights to knowledge.

Innovation therefore is the major step in the development of technology without which technology remains just an idea.

Furthermore, innovation - to be successful - and in addition to the technology development process, must include a business plan which defines the potential implementation, introduction, marketing, and financing of the technology subject to the innovation. Otherwise, innovation may suffer under lack of effective direction, or may lead to technological developments which are not usable or marketable.

With an increasing rate of technological obsolescence, technological forecasting and analysis to discover technological voids and opportunities, as well as the relative technological position, has become one of the most important responsibilities of decision and policy makers. This includes evaluation of the dynamics of the environment in which technology is developed and a comparison with the existing environment.

1.2 FACTORS INFLUENCING TECHNOLOGICAL CHANGE

The factors influencing decisions that affect technological change depend on the current technological position, product orderbook, product marketing potential, competing industry position and capability (including relative costs), input factor prices, available technological options, and trends in both product technology and process technology.

Figure 1.1 is a simple diagram in which the relations among the various factors influencing technological change decisions are shown. Usually the most important factors are relative input prices, productivity in terms of output per unit input or cost of product, and ability of the process, service, or product to meet changing requirements or demands including such characteristics as quality. The most important decision variables in the management of technological change are as discussed:

1. technology choice;
2. method of innovation, adoption, development, and introduction of new technology;
3. timing of adoption, development, and introduction of technological change;
4. rate of introduction and use of technological change; and,
5. level of investment into technological change and resulting capacity introduced.

In most technology change decisions, only some of the above management decision variables apply. There are many alternative sources for technological change. In many industries the primary source is usually research and development (R&D) performed in-house or under contract.

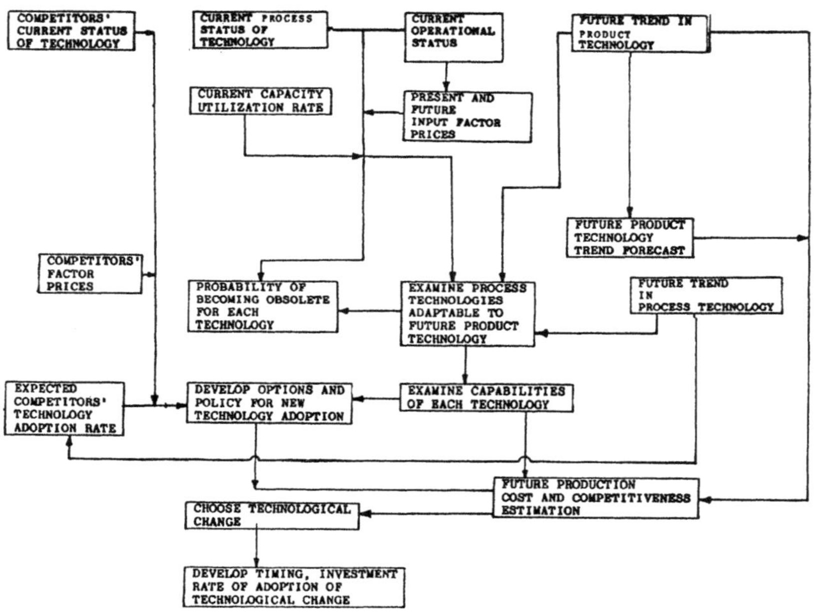

Figure 1.1. Factors Influencing Technological Change.

The impact of technological change on productivity can be pervasive if effectively timed, managed, and applied. On the other hand, technological change may cause no more than a ripple if ill-timed, badly manager, or ineffectively applied. Technological change can be achieved by transfer or purchase of new product, service, or process technology, adoption or spin-off of new technology, or in-house development through research and development. In many countries, detailed strategic plans are developed to guide decisions on timing, development, transfer, and adoption of technological change. Such planning usually involves economic analysis of the advantages of alternative methods of technology acquisition. It includes trade-offs of probability of success in research and development aimed at technological change in terms of acquiring new technology in a timely fashion at a competitive cost, as well as the probability of advancing technology beyond the development achieved by competitors or elsewhere.

It is interesting to note that some of the most advanced industrial countries in the world are purchasing significant process and product technology while concentrating their in-house or domestic technology development on very narrow or specific technological issues, based on the discovery of critical technological voids or opportunities. The reason for this tactic is the recognition that adoption of new technology often requires development of new interface technology to permit effective use of the new technology.

In some countries, research and development is sometimes involved in duplicating technological changes or advances made elsewhere with little evaluation of the trade-offs between domestic technological advances and transfer or import of the technology from the point of view of economic and financial cost, timing of technological change, and competitive aspects. The reasons for this approach are often national pride, lack of unbiased judgement, or others. Another issue is the discovery of technological innovations in other, often unrelated, industries with potential applications. Here again, other sectors of the economy appear to be more alert to such technological transfers and devote more substantial resources to the discovery of such opportunities and applied research into technology transfer.

A most important issue is that technological change must not just happen and - in particular - only happen in response to competitive or market pressures, but must be planned. The most successful firms consider technological change a strategic issue which requires medium- to long-term planning towards accomplishment of a strategic goal. The sources of technological innovation generally fall into the following major categories:

1. in-house research and development;
2. external research and development specifically oriented to technology improvements;
3. other firms (domestic or foreign);
4. other industries using or developing technology of potential use;
5. equipment manufacturers;
6. basic research organizations such as government laboratories, universities, public/ private research organizations, and more;
7. product technology in use with potential application; and,
8. other sources.

Decisions relating to technological change are affected by the competitive position of the firm in terms of costs per unit output which are, at least in part, affected by the level of technology used and of input factor prices. While the level of technology used will affect the productivity of input factors such as investment capital and labor, there are other factors which will affect the productivity of input factors as well. The cost of acquisition of technology from different sources usually varies widely. Similarly, the time of introduction of new technology will be influenced by the choice or source of technology and method of acquisition or transfer. While research and development may cost substantially more in time and money than the purchase of the same technology from external sources, in-house development may provide more effective integration with on-going processes and procedures, as well as permit custom development of new technology. This may sometimes - but not always - justify internal technology development.

1.3 TECHNOLOGICAL VOIDS AND TECHNOLOGICAL DEVELOPMENTS

Industries are usually divided into mature and innovative, with other subdivisions defined in between. Mature industries are generally assumed to have reached their peak and to coast along on their past technological developments with an inevitable downturn in future. On the other hand, innovative industries are assumed to be up and coming, based on the recognition of vast voids in technology and technological opportunities which, if filled, would satisfy a real and valuable need, and therefore result in effective growth. Mature industries, on the other hand, are usually assumed to have filled most technological voids in their field and, as a result, have few, if any, technological opportunities. Technological voids are usually identified through:

1. theoretical study and limit analysis;
2. recognition of an inability to meet an existing or perceived service demand;
3. discovery of new opportunities which require technological development or advance;
4. competing developments or development of competition;
5. discovery of technological developments in other fields with potential application; and,
6. technological forecasting.

The process of technological void recognition and technological development is quite complex and involves various factors which lead to the development of technological advances. Voids are sometimes recognized by study of other and often unrelated areas of technology. For example, laser technology developed initially for the improvement of physical measurement was subsequently recognized to provide useful advances in medical, printing, and communication technology.

1.4 ACQUISITION OF NEW TECHNOLOGY

One of the major driving forces in industrial development has been the successful approach to the acquisition of new technology. In the past, economists did not associate technological development with economic growth because they interpreted technological change as being closely tied to innovation and innovation with lengthy, expensive research processes. Technological change often involves purchase and subsequent transfer of new technology, together with technical assistance in its use. Decisions affecting technology acquisition require:

a. the evaluation of technology adaptation and implementation capability. In other words how and in what manner can new technology be adopted or introduced?
b. the setting of technology acceptability standards and constraints;
c. the determination of training and skill development requirements for effective introduction of the new technology.

Usually there exists both institutional constraints and incentives to any new technology introduction. These may consist of political, cultural, regulatory, economic, demographical, and other constraints and incentives. The evaluation or assessment of the effectiveness of a technology transfer profess is usually based on a value system for technology transfer assessment which uses effectiveness standards.

The primary objectives of technology and technology transfer assessment are to identify and analyze the relevant economic, technological, legal, institutional, and environmental consequences of projected technological changes; to analyze the ability to accommodate the technological change; to compare the alternative technological choices available; and to identify and analyze the uncertainties and risks associated with alternative technological choices, and methods of introduction of new technology.

Problems in the transfer of technology usually involve

1. maintaining technology transfer flow,
2. timing of introduction and updating of technology,
3. effectiveness of the technology transfer,
4. decisions regarding stepwise versus continuous technology transfer, and
5. feedback and feedforward of technology use information, particularly relative to changes in technology or its use in the new environment.

Technological decisions - once made - must be implemented without delay. To choose technology effectively, a large amount of information, extensive experience, good judgement, and an ability to make rapid decisions and commitments are required.

Technological change can be generated or triggered by different processes. It can originate from internal research and development results, an internal process of innovation, new technology acquired from external sources, such as competing firms, vendors of technology, universities, and other research and development organizations. It can be transferred from other industrial sectors by a process of adaptation. Technological transfer is either (1) vertical - from more general to more specific, or (2) horizontal - from one application to another. The rate of technological change is increasing at an alarming rate.

Appropriate technology transfer policies usually vary according to the stage of development of a country and industry, its technological absorption capacity, and its own objectives. The issues of concern in judging the appropriateness of technology are usually

- cost of transfer
- appropriateness of technology transfer
- effects of technology transfer on learning, and
- technological development and technological dependence.

Technology has many characteristics of a public good such that the marginal cost of communicating it is usually very low compared with the initial cost of development or acquisition. As the technology market is highly imperfect, the price charged for technology tends to be oligopolistic. The effectiveness of technology change is usually measured in terms of:

1. changes in output (capacity, quality, amount of scrap, etc.);
2. changes in inputs (manhours, machine hours, material inputs, preparation, programming, design, planning, etc.);
3. effects on other activities (adjacent processes, transfer, storage, product flow, etc.);
4. effects on the environment; and,
5. adaptability (worker acceptance, skill acquisition, regulatory acceptability, etc.).

Other factors influencing the success of introduction of new technology are management decision-making ability, government cooperation, corporate structure and infrastructure, and more.

The most important factor is probably the effect of maturity of an industry subject to a low rate of product, market, as well as process technology change on the management of technological change. Similarly, competitive pressures in such an industry usually chance slowly as do their effects on technological change requirements. As a result, mature industries invest little effort in the development of technological change, and their management of technological change is largely reactionary.

1.5 MANAGEMENT OF TECHNOLOGICAL CHANGE

The effectiveness of technological change depends on many factors. For example, the choice of new process technology must (1) fit the projected product to be produced by the process; (2) permit efficient use of available sources of inputs such as materials, services, and labor; (3) allow effective integration into existing production and assembly approaches; and, (4) must be environmentally acceptable. Similarly, timing of the introduction of new process technology must be appropriate to (1) the level of development and diffusion of the new technology, (2) capacity needs of the plant, (3) competitive position of the plant in terms of production costs, and (4) rate of obsolescence of the technology.

The size and rate of introduction of new technology again depend on (1) capacity requirements, (2) investment capability, and (3) acceptability of the new technology, including the rate of training of technology operators. Finally, the utilization of new process technology depends on (1) required output, (2) acceptance by labor, and (3) relative cost of production, and more.

While it is generally recognized that technological change is critical to any firm's success and growth, and that any change in process capacity and capability usually introduces some technological change, the management of technological change remains among the least understood aspects of operations/production management. The contradiction is that technological change at the time is credited as a major contributor to the advance of any firm as well as the economic growth of regions and nations.

The study of the factors which contribute to the effectiveness of management of technological change at the process level is a problem which has received scant attention. The reason is probably the lack of availability of adequate data on process performance under conditions of technological change. Similarly, the theoretical basis developed for the analysis of exogenous technological change at the aggregate level or endogenous technological change at the level of the firm considered technological change a strategic decision concerned with the longer term impact on the firm or nation. Another reason may be the comparatively long time between technological changes at the process level in the past, which allowed consideration of process performance changes to be attributed mainly to learning. As technological change at the process level was either incremental or occurred at very long intervals, performance at the process level could not be effectively used to explain the impact of different approaches to the management of technological change on the success of the technological change.

In recent years changes in process capacity or process replacement usually implied technological change. There are a few processes today which do not undergo significant technological improvements even in a matter of a few years.

This provides an opportunity for the study of the management of technological change at the process level by analyzing process performance and identifying the effect of technological change on changes in process performance.

A related problem is the determination of the way in which technological change affects process performance and the identification of the management decision variables which influence the effectiveness of technological change in improving process performance. Comparative analysis of different models of management of technological change is designed to identify effective tactics for management of technological change.

Technological change at the process level can be introduced incrementally or radically. The rate of introduction depends on process capacity change requirements which, in turn, depend on:

1. market developments and prospects;
2. competitive position;
3. relative costs;
4. utilization of existing process capacity;
5. barriers to technological change at the process level;
6. availability of new process technology; and,
7. investment capability.

The management of technological change involves the identification of a need for technological change and availability of new technology. The need for technological change is usually determined by comparing the marginal improvement in productivity of the process in use with changes in output price (or profit levels) as well as input (i.e., labor) costs. Large price reductions or labor cost increases may demand improved productivity increases. Once the need is established, a choice from among available new process technologies must be made. This choice will be affected by

1. the start-up productivity of the new technology,
2. the expected slope of its experience curve, and
3. the availability and reliability of the new process technology.

Another management decision is the timing of introduction of the new technology and related decisions, such as lead time and the bundling, to be used in the acquisition of the new technology.

Similarly, management must decide how to transfer output from the old to the new process technology. If new process capacity is equal or larger than that of the old process, then obviously all the output could be transferred quickly and productivity improved, based on the productivity of the new technology. As we will note, there are many alternative strategies available for transfer of output from old to new process technology.

The choice, timing, scale, and rate of introduction of technological change are interdependent as shown in Figure 1.2 and depend on both exogenous and endogenous factors. Management usually controls some - but not all - factors and, if effective, attempts to maximize the benefits from technological change. Management is responsible for the choice, timing, scale, rate of introduction, and finally utilization of the new process technology. But there are many constraints on management in this decision making process. The choice of technological change, for example, is influenced by new process technology such as:

1. availability,

2. state of development,
3. operational, product, and productivity advantage,
4. acceptability by labor and staff,
5. terms and conditions of acquisition,
6. regulatory constraints (OSHA, etc.),
7. investment requirements and capital availability,
8. minimum/maximum and optimum size,
9. environmental constraints, and
10. rate of obsolescence.

In part the choice affects the timing of technology acquisition which, in turn, affects the scale and rate of introduction of the new technology.

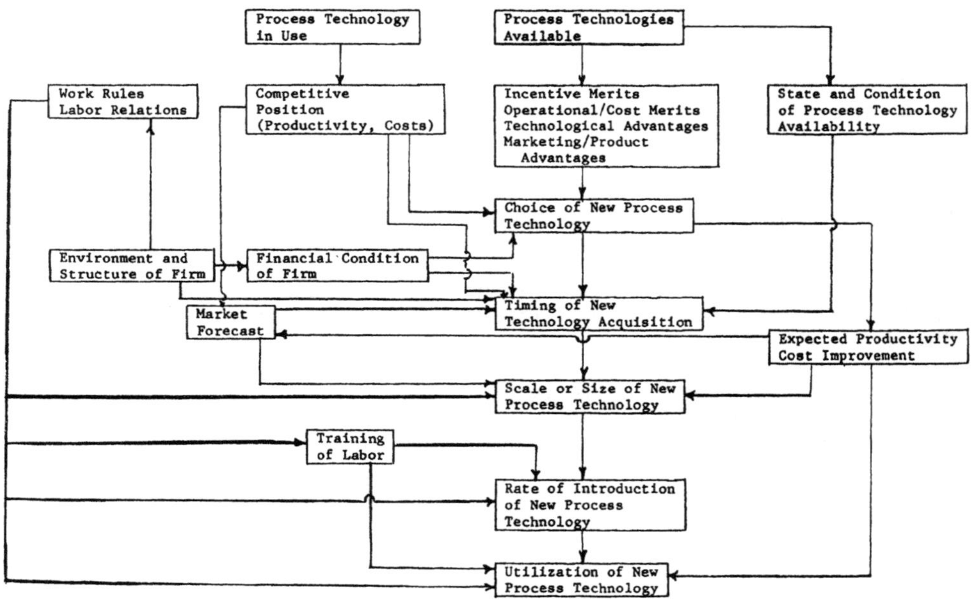

Figure 1.2. Relations of Choice, Timing, Scale, and Rate of Introduction of Technological Change.

Technological change is driven by the need for improvement in process productivity (cost), reliability and output quality, as well as capacity. Technological change can therefore be cost (competition) or demand driven, though in most cases today the incentives for technological change are usually derived from both a desire to improve productivity (costs) and increase capacity (Figure 1.3). In most cases, capacity increase (or replacement) today involves some technological change.

1.6 MODELLING IN THE MANAGEMENT OF TECHNOLOGICAL CHANGE

A model of the management of technological change is shown in Figure 1.4. it indicates the decision making processes required. In other words, the management of technological change

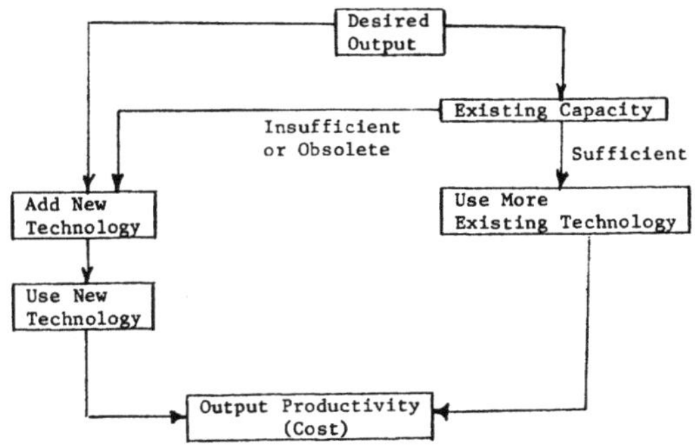

Figure 1.3. Transfer of Outputs from Old to New Technology.

consists of a number of partially interdependent management decisions which determine the characteristics and process of technological change. The results of the analysis of the effects of technological change determined quantitatively for each process at each firm would be used to identify the contribution of the various management decision variables on the effectiveness of the technological change. Similarly, the choice or use of decision variables by the different firms (such as timing, bundling, training, lead time, etc.) form the qualitative parts of such a model.

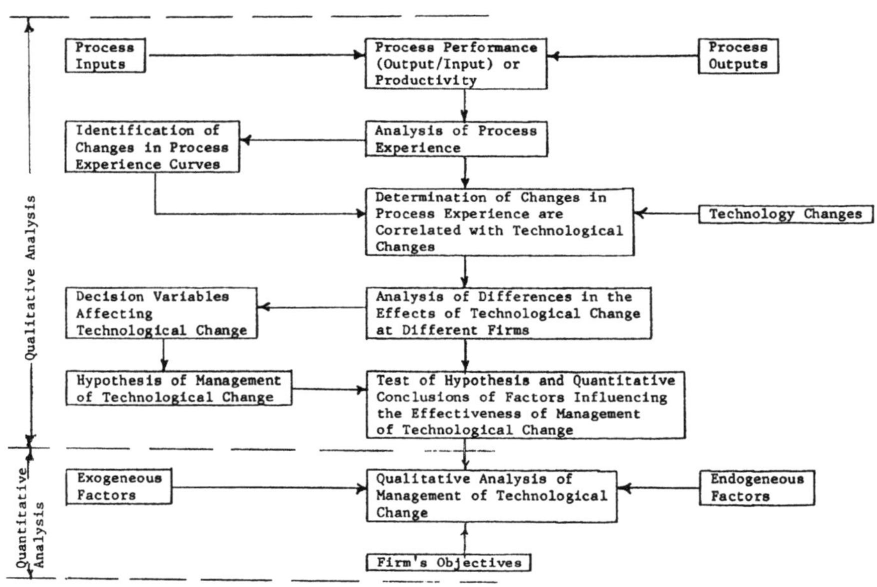

Figure 1.4. General Framework of Model.

Therefore such a model consists of a statistical analysis of process performance under conditions of technological change and provides the data for the analysis of the effectiveness of the management of technological change. The use of management decision variables or the strategy used in managing technological change can then be analyzed qualitatively by evaluating exogenous and endogenous factors which influence the choice and value of the decision variables used by management (Figure 1.5).

1.7 THE TECHNOLOGICAL CHANGE PROCESS

Technological change is usually the result of an invention caused by conception and supportive research if required, a subsequent innovation process in which the idea is not only refined but translated or developed into a useful product or process with potential applications. The actual technological change, usually the result of commercialization of the invention, takes place during the technology innovation process.

The time from invention to innovation and length of the innovation process can be quite large. The length of this gestation period depends on many factors such as

1. visibility of the invention or innovation,
2. protection applied or employed,
3. relevance of invention or innovation,
4. commercializability,
5. range of possible uses or applications,
6. cost of bringing invention or innovation to commercial state,
7. expertise required, and
8. maturity and effectiveness of existing, competing technologies.

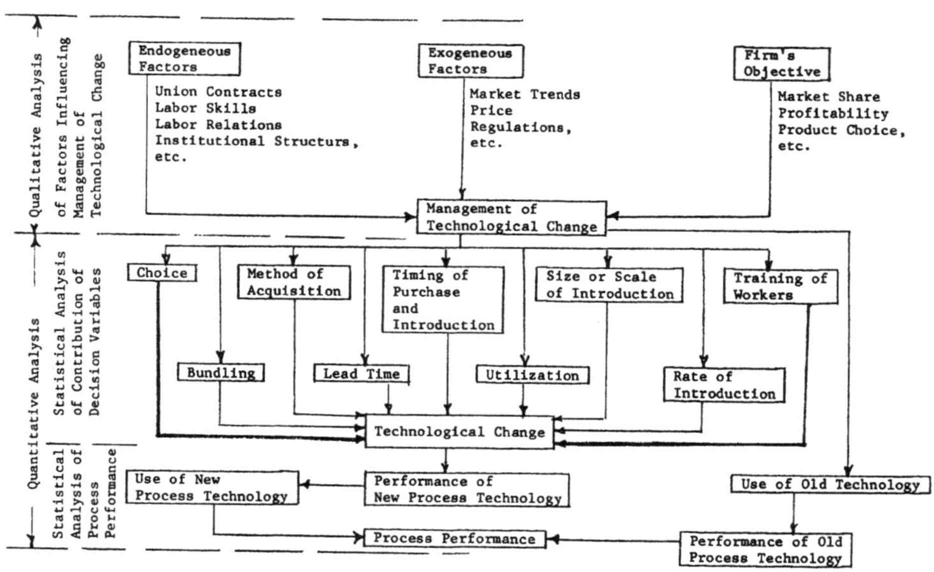

Figure 1.5. Diagramatic of Relationships of Factors of Management of Technological Change.

When a new technology is introduced, existing technologies and their users affected by it will usually react, often in an attempt to prevent the new technology from gaining a foothold. As start-up or innovation process costs of new technologies are usually high and start-up of performance of new technologies, often non-competitive with that of existing technologies, such action can cause delay or abortion of the introduction of new technologies.

The technological change decision process must consider such and many other factors, both exogenous and endogenous as shown in Figure 1.6. Some of these can be readily determined or at least estimated by technological change decisionmakers, while others are difficult to obtain. Various estimating and statistical techniques should then be employed to assure the highest level of validity of inputs or assumptions in the analysis performed to support technology change management decisions. For example, competitors' costs and pricing strategy could be estimated from historic price data, estimates of competitors' costs based on location, sources of input, and similar factors.

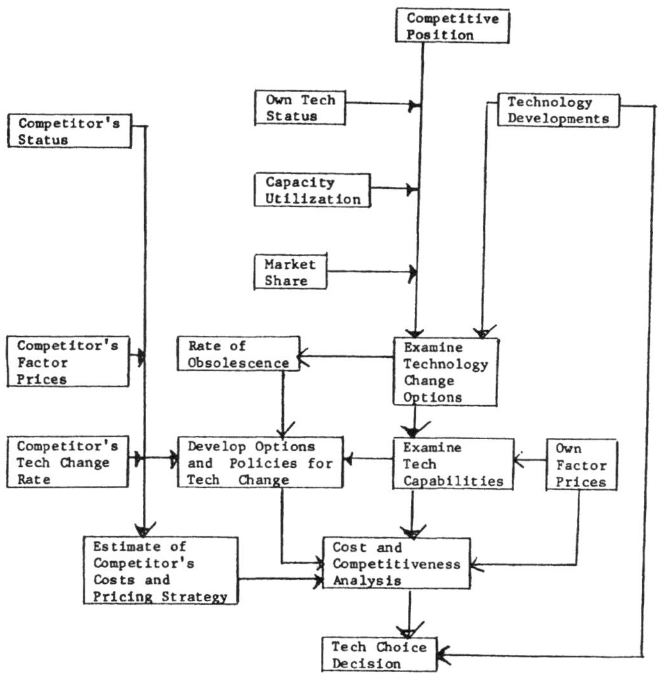

Figure 1.6. The Technological Change Decision Process.

As noted before, the choice of new technology is only one of several important decision variables. Equally, if not more, important are the decisions of the timing, scale, and rate of introduction (or utilization) of the new technology. There are various related issues affecting the technological change or choice decision such as:

1. the source of the new technology, its ownership, nationality of owner, and the environment in which the new technology was developed;

2. the state or stage of development of the new technology. Is it a new invention (discovery), an innovation, or a commercially available process or product?
3. the type and degree of protection of the innovation/invention or process/product. Is it patented? Can it be licensed, sold, etc.?
4. What methods are available or could be developed for the transfer of the new technology?
5. Are there substitute technologies available or under development?
6. Are the economics of the new technology attractive?
7. Is the new technology introducible? Will it require changes in skills, attitudes, philosophy, labor, public relations, and more?
8. Externalities.

Therefore, we must remember that a technological change (choice) decision is complex and requires consideration of many issues which are discussed later in this book. It is also a dynamic process in terms of the performance of the old and new technology. Both will continue to react with time and usually improve their performance as shown in the case of an old and new process technology (Figure 1.7).

Obviously there are often new processes or technologies which do not undertake the performance (productivity) of the old process technology but start with better performance than the old technology can provide and continues to improve on it at a higher rate. In that case, the performance (cost or productivity) curves of the old and new (process) technologies never cross. In some cases, a new technology is also capable of performing tasks that the old technology could not accomplish.

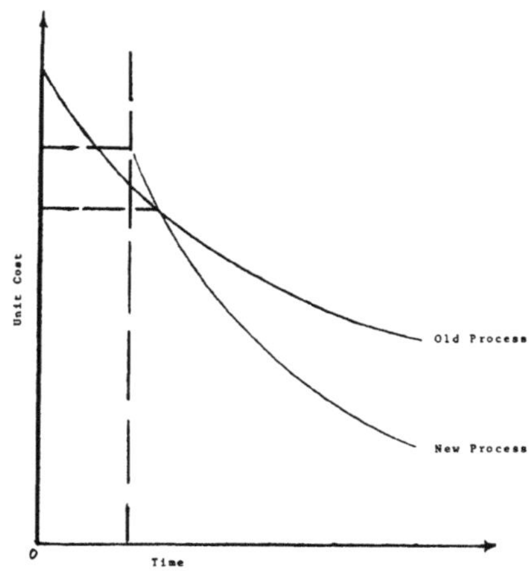

Figure 1.7. Productivity (Performance) of Old and New Process Technology.

1.7.1 *Technological Change Process Stages.* The technological change process is usually divided into several stages, such as

Stage 1 - Invention or Discovery

The basic idea is discovered by chance or research.

Stage 2 - Innovation or Development

The idea is developed into an effective process or product.

Stage 3 - Diffusion and Commercialization

The process or product is adapted to production or user needs and commercialized, marketed, used, or diffused among various users.

The time required for each stage and time between stages varies widely with source of invention or discovery, ownership, method of invention or discovery and innovation used, type of idea, sophistication of environment, and more. The time between invention or discovery and innovation is often called the gestation period.

Gestation periods vary widely. It was 9 years between the discovery of radar in 1925 and its first application (innovation) in 1936. DDT, on the other hand, had a gestation period of 62 years from 1874 to 1942. The gestation period is largely affected by the recognition of the commercialization potentials of the invention, financial ability of supporter of the innovation stage, and the absence of major or powerful competitor's or competing processes or products. Other factors such as visibility, relevance, range of uses, expected cost of innovation, required expertise in innovation, and maturity of existing competing technologies also affect the length of the gestation period. In fact, there are many inventions or discoveries which are never subjected to innovation and never attain commercialization because they fail to satisfy one or more of these important factors.

1.8 COMPARATIVE MANAGEMENT OF TECHNOLOGICAL CHANGE

There are often significant differences in the effectiveness of the management of technological change. This may be the result of differences in the choice of management decision variables which affect technological change. This may be the result of differences in the choice of management decision variables which affect technological change. Differences in choice of management decision variables may be the result of cultural, social, organizational, or other factors. Comparative management is the study of the management phenomena on a comparative basis which has as its objective the detection, identification, classification, measurement, and interpretation of similarities and differences among management phenomena being compared.

The management phenomena of interest are the management decision processes, techniques, value systems, and other observable phenomena which affect the management of technological change. Organizational, technological, market, financial, and environmental factors, among others, can usually be identified as influencing management effectiveness as shown by the comparative technological change management constructed in Figure 1.8.

Using an ex-ante set of presumptions as to how management identifies issues and communicates, accepts, and implements decisions, an analysis of the results of a qualitative analysis of the management organization, environment, and procedures can sometimes bs used to determine if information and performances are representative of the values, attitudes, and behavioral responses expected. The result of a comparative study should provide explanations of the differences in the effectiveness of the management of technological change as well as a test of the more general comparative management hypothesis established by previous relevant research into management approaches.

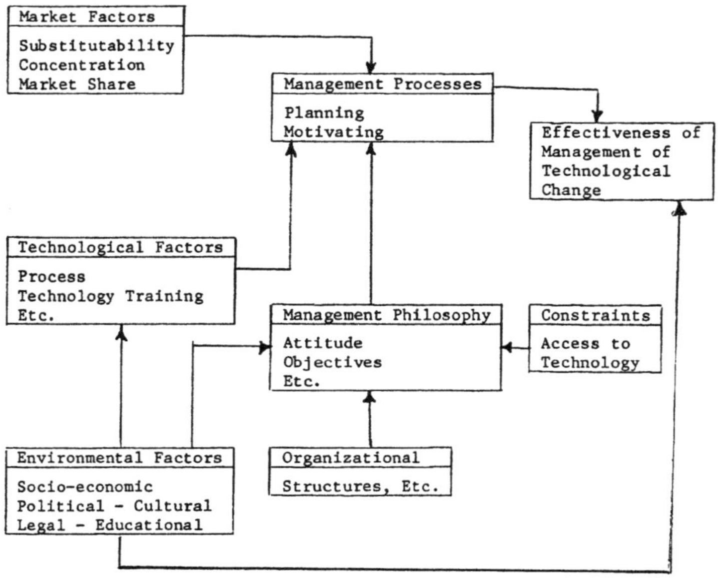

Figure 1.8. Construct of Comparative Management of Technological Change.

Differences in labor attitudes and motivation, management values, and other determinants of leadership and decisionmaking must often be studied to explain differences in effectiveness, if any, in management decisions affecting technological change. Such theoretical models and various management behavior theories (such as the contingency and path-goal theory) provide useful analytical frameworks for analysis.

References

1. Mansfield, E., "The Economics of Technological Change", W. W. Norton & Co. Inc., New York, 1968.

2. Kekala, P. J., "Iron Ore - Energy, Labor, Labor and Capital Changes with Technology", J. Wiley, New York and Boston, 1973.

3. Enos, J. L., "Petroleum Progress and Profits - A History of Progress Innovation", The MIT Press, 1962.

4. Hollander, S., "The Sources of Increased Efficiency: A Study of DuPont Rayon Plant", The MIT Press, 1967.

5. Griffin, J. M., "Long Run Production Modelling - Electric Power Generation", Bell Journal of Economics, March 1982.

Bibliography

Bavid, R. N., "Production Functions Productivity and Technological Change", Working Paper No. 65, Research Program in Industrial Economics. Cleveland, Case Western Reserve University, August 1975.

Barnett, H. G., Innovation: The Basics of Cultural Change, McGraw-Hill, New York, 1953.

Bright, James R. (Ed.), Research, Development, and Technological Innovation, Richard D. Irwin, Homewood, IL, 1964.

Bronfenbrenner, Martin and Douglas, P. H., "Cross-Section Studies in the Cobb-Douglas Function", Journals of Political Economy, Vol. 47, December 1939.

Burns, Tom and Stalker, G. M., The Management of Innovation, Tavistock Publications, London, 1961.

Chandler, Alfred D., Jr., Strategy and Structures, The M.I.T. Press, Cambridge, MA, 1962.

Converse, Paul D., "Marketing Innovations: Inventions, Techniques, Institutions", in F. E. Webster, Jr., (Ed.), New Directions in Marketing, American Marketing Association, Chicago, IL, 1965.

Cooper, A. et al, "Strategic Responses to Technological Threats", Academy of Management, Proceedings, 1973.

Emery, F. E. and Trist, E. L., "The Causal Texture of Organizational Environments", Human Relations, pp. 21-31, February 1965.

Engle, J. D. and Blackwell, R., Consumer Behavior, 2nd Edition, 1973.

Fisher, J. C. and Pry, P. H., "A Simple Substitution Model of Technological Change", Technological Forecast. Soc. Change, 3, pp. 75-88, American Elsevier Publishing Company, New York, 1971.

Gold, B., "Economic Effects of Technological Innovations", Management Science, September 1964.

Gold, B., "Research, Technological Change, and Economic Analysis: A Critical Evaluation of Prevailing Approaches", Quarterly Review of Economics and Business, Spring 1977.

Gold, B., Technological Change: Economics, Management, and Environment, Pergamon Press, Oxford, 1975.

Hagerstrand, T., Innovation Diffusion as a Spatial Process, Chicago University Press, 1967.

Hawes, D. K., "An Inspection of Innovations", unpublished paper submitted for a graduate course in consumer behavior, Ohio State University, Columbus, OH, 1968.

Hollander, S., "Sources of Increased Efficiency", W. Irvin, New York, 1968.

Kakela, P. J., "Iron Ore: Energy, Labor, and Capital Changes with Technology", J. Wiley, New York and Boston, 1973.

Kendrick, J. W. and Vaccara, B. N., "New Developments in Productivity Measurement and Analysis", University of Chicago Press, IL, 1980.

Little, A. D., "Patterns and Problems of Technical Innovation in American Industry", The Role and Effect of Technology in the Nation's Economy, Hearing before Congress, 1980.

Mansfield, E., "The Economics of Technological Change", W. W. Norton & Co., New York, 1968.

Melcher, A. J., "A General Index of Technological Change", ORSA/TIMS, 1977.

Sato, R. and Suzawa, G. S., "Research and Productivity - Endogenous Technical Change", Auburn House Publishing Co., Boston, 1983.

Sverdrup, C. F., "Considerations Regarding Improved Productivity Based Upon Experience of Series Production of Merchant Ships", IREAPS, International Shipbuilding Symposium, 1982.

Szekely, J., "Toward Radical Change in Steelmaking", Technology Review, 1979.

Walters, A. A., "Production and Cost Functions: An Econometric Survey", Econometrica, Vol. 31, pp. 1-66, January/April 1963.

Ward, B., "Robotics and Reality", SKY, November 1982.

Woodward, Joan, Industrial Organization: Theory and Practice, Oxford University Press, Fair Lawn, NJ, 1965.

2. Research in the Management of Technological Change

Management of technological change has been the subject of research for over 30 years. Technological change has been studied by social and political scientists largely concerned with the organization and mechanism of innovation and the problems of diffusion of technological change. Sociologists have researched the spread of ideas and information and the impact of technological change on society and individuals. Social geographers were interested in the spatial issues involved in technological diffusion. Economists, on the other hand, studied the causes and effects of technological change with particular attention to their quantification with a view towards establishing methods for the determination of the economic impact of technological change. Systems and management analysts researched the process of technological change and, together with economists, have studied the relations between technological change and productivity or productivity growth. Organizational scientists are interested in the effects of technological change on organizational structure as well as on the role of management and the individual worker involved in the changed process or consumer using a changed product.

The effect of technological change on plant or process performance has been studied extensively, as has the process of technological change. Early technological change studies assumed technological change to be an aggregate phenomenon which accounted for all performance changes. Recognition of the limitation of such exogenous global approaches has led to an increasing use of microeconomic analysis to explain the impacts of technological change.

Studies of the costs involved in technological change and the derivation of empirical counterparts to short-run static cost functions, used in microeconomic theory has greatly advanced the understanding of the role of technological change and its impact on economic growth, as first explained by Solow [6].

The management of technological change involves the process of technological change as affected by internal and external factors as well as the effect of technological change on the firm, process, or product and its environment. It also involves the comparative management of technological change and its impact on productivity. This requires effective measurement of productivity. Research in productivity measurement often involved microeconomics because this approach permitted the establishment of implicit relationships between costs and outputs. Production and cost functions are therefore often used in technological impact studies.

Research of the various stages in the process of technological change from invention and discovery to the various stages of innovation to implementation and improvement of technology diffusion is now performed by many investigators.

In recent years experience curves, a popular tool for the analysis and forecasting of process performance in the seventies, have become somewhat discredited because of the unreliability of this approach to the measurement of improvements introduced by a process technology under some conditions of applications. The approach, though, provides an effective means for analyzing changes in process performance and is now popularly used. Many researchers have suggested that the use of a static experience curve which assumes no change to the process in terms of capacity or technology was unrealistic in today's environment. Some earlier investigators also proposed the use of a disaggregate approach to the analysis of the effects of technological change at the process level, to obtain a better understanding of the factors which make for better management of technological change.

Yet experience curves provide an effective means to evaluate competitive interaction of process technologies within a plant and competitive conditions between different plants. Experience curves similarly can be used to study the effects of different strategies of choice, introduction, and use of new process technology on the effectiveness of technological change. They provide a good method for the study of the transitional effects and the dynamics of change. In turn, this can give meaningful insights into the impact of different approaches to the management of technological change. Experience curve analysis is therefore presented as a method for the analysis of factors which influence the effectiveness of technological change. This literature review covers research into factors of productivity growth at the industry or plant level, research into the use of experience curves, and research in comparative management - all areas of investigation of importance in the understanding of the management of technological change.

2.1 INDUSTRY STUDIES

Mansfield [1] did the pioneering work in quantitative analysis of the management of technological change by examining the rate of diffusion of technological change among companies - in particular, industries such as bituminous coal, iron and steel, brewing, and railways. His results indicated that after a new technology is adopted by one or two companies it spreads to others, initially at a slow rate which afterwards accelerates, levelling off again as most companies in the industry adopted the new technology. Subsequent studies of other industries confirmed the S-shaped curve of technological diffusion, advocated by Mansfield [1].

Other studies of technological change at the industry level were performed by Kekala [2] who evaluated the technological development of the steel industry and concluded that technological change was the principal factor causing productivity improvements. Similar conclusions were reached by Enos [3] who studied the oil refining industry from the point of view of the historic contributions of changing technology. Hollander [4] used a microeconomic approach to study the impact of technological change on rayon manufacture. Other researchers such as Griffin [5] investigated the effect of technological developments on electric power generation.

The approach taken by these and other investigators varied from historic technology development analysis and empirical study of the impact of technological change to microeconomic studies of firms in the industry. In all these industry studies, technological change was assumed to be endogenous and the effects of environmental or external effects were ignored. Similarly multinational and intra- or inter-industry factors were not considered.

Many of these studies were encouraged by the conclusion of the macroeconomic research of Solow [6] and others in the late 1950s, which showed definitive links between technological change and productivity, but concluded that the concept of embodiment of technological change may be restrictive, and that an aggregate approach to a study of technological change may ignore many important aspects, such as the separation of the impact of investment versus that of learning, and the effects of the rate of introduction of new technology. As a consequence, research in microeconomic aspects of technological change attracted many investigators such as Hollander [4] after 1960. One industry study which carried Hollanders' microeconomic approach of technological change further by introducing rates of return effects under endogenous technological change was that of Sato [7] who developed a model for the chemical industry.

Issues of environmental and other external impacts on the effectiveness of technological change have not been studied explicitly, although extensive research in comparative management and management of multinational firm address this problem indirectly. There are a number of studies on the management of the firm, such as those by Thompson [8] and Lawrence and Lorsch [9], which address the problem of external factors and their effect on managements' approach to technological change. Similarly, researchers such as Sethi [10] developed general models of management response to external changes under condition of technical change. Yet most of these studies do not identify the particular management factors and decision variables which influence the effectiveness of technological change. Exception to this are the studies reported by Abernathy [11].

2.2 EXPERIENCE AND TECHNOLOGY CHANGE

Most technological change in production processes is evolutionary and consists of the replacement by or addition of more advanced process capacity. Wright [12] drew attention to the concept of learning curves as early as 1936 by describing how direct labor costs decreased with experience; research in the application of learning curves to the management of production was first reported by Andress [13] in 1954. He investigated use of learning curves in the planning of production capacity and performed a number of empirical studies to establish the validity of learning curve characteristics. Subsequently, extensive research on the effect of learning on productivity in the manufacturing industry was performed by Holdham [14] who used extensive empirical data to prove the validity of this approach.

Others like Hirshman [15] contributed to the development of the theory that evolution of technology is governed by accumulated experience. They showed that the theory holds for diverse industries and that the experience curve reflects the combined effects of a multiplicity of factors such as growth of worker skill, improvements in equipment performance, better process management, and more. In other words, that experience plays a vital role in the evolution of productivity. This was amplified by others who found learning significant in the use of existing process technology.

Since the Boston Consulting Group [16] published its pioneering book entitled "Perspectives on Experience" in 1968, experience curves have become an important tool in predicting future productivity or costs of a process. The concept of an experience or learning curve is intuitively appealing and empirically testable, but it has limitations. For example, it assumes that additions to or replacements of process capacity will exhibit the same learning curve as the original process and will have an identical productivity or cost as the original process has at the time of their introduction. Abernathy and Wayne [17], Sallenave [18], and Kiechel [19] point to some of the limits of the experience curve with particular reference to the effects of time, the fact that no process technology can improve forever, and the differences between the experience gained over time and with respect to cumulative output.

The experience curve approach is generally used as a basis for making strategic production decisions on quantity and rate of production and to forecast future production costs. It is generally assumed that productivity or costs follow the established experience curve, even when massive capacity additions are required and introduced. This basically implies that capacity additions, whenever made to a process, will provide identical technology as the original process with the same maturity as achieved by the original process, and that increased capacity as well as larger scale production will not affect the rate of productivity growth. These assumptions are unrealistic in all but a few static processes or processes in which technological and capacity changes are very gradual indeed.

Some recent authors considered the effects of joint production by diverse processes on the experience curve. Others like Womer [20] dealt with a model of production on orders, where the order size and production rate each influence the process productivity or cost of production. The use of the experience curve in strategic planning is discussed by Hedley [21] and Ghemawat [22].

The effects of experience on the cost of production, in the presence of learning, when optimizing production rate, were studied by Washburn [23] who developed a model which adjusts production rate so as to maximize discounted costs or profits over a production planning horizon.

The experience curve also offers an effective way to model the effectiveness of management of technological change as it permits the representation and study of the transition from an old process technology to a new process technology.

One of the problems vexing many of the researchers in learning curve effects was the fact that only part of the productivity or cost improvements represented by the learning curve could be explained by learning, where, as noted above, learning included both improvement in worker performance with experience as well as improvement in the use of the process involved. The remainder appeared to be caused by technological improvements resulting from (1) incremental improvements in the process itself, (2) gradual addition of increasingly advanced process technology; (3) technological adjustments or refinements to the process; and, (4) radical technology change.

The authors of papers on experience curves generally assumed that experience resulted in an exponential improvement of productivity (or cost) with output which continued indefinitely. Such exponential experience curve characteristics result in an ever declining percentage of productivity improvement which soon become insignificant. As a result, few processes operate along their experience curve for a long period of time or a large amount of cumulative output before a technological change in or abolishment of the process occurs.

While most empirical results at the process and industry level proved the validity of the exponential experience curve, the tests were usually performed over comparatively short periods of time or small quantities of cumulative output. Besides this, the rate of technological change in the past was much slower and few technological improvements became available during the short time horizons considered in many of these studies.

In recent years various researchers concluded that accumulated experience contributes to productivity not only by learning in the traditional sense or learning by diffusion, but also by the scale and rate of introduction of new capacity (or technology), utilization of existing capacity, and environmental factors. Change in scale in process often substitutes for change in process technology and may offer greater improvements in productivity, particularly as it is often easier to change process technology in relation to scale than it is to change scale in relation to process technology. A typical example of this factor is the use of increasingly large tankers in oil transport. While this confirms the hypothesis of the contribution of scale, it fails to explain all experience gains. The reason may be that the hypothesis was usually considered in isolation. In reality, it seems reasonable to assume that significant interactions may be involved between the experience effects of scale, utilization, etc., in addition to learning by doing and diffusion.

Another issue is that conditions have changed radically in recent years. Few process technologies are able to perform effectively for the whole of their operating life because better process technology is usually available long before then. While the ratio of economic to operating life was close to one for many processes until a few years ago and equipment was often used until nearly worn out, this ratio is now only 20-40% for most process equipment which is, as a result, often discarded long before the end of its operating life.

The experience curve therefore now assumes a new role. It can be used to model not only the learning achieved in the use of a process but the transition as the process goes through an evolution of technological improvements.

Experience curves have recently been shown to be unreliable at times, as a planning and control tool as shown by Sallenave [18]. There is mounting evidence though that the increasing lack of reliability of learning or experience curves in predicting process productivity and costs is mainly due to gradual or radical technological changes in the process itself. The experience curve can no longer be considered a static statement which projects future process performance. To be effective, it must become a dynamic concept, subject to the effectiveness of management of technological change.

This can be accomplished if the learning or experience curve is assumed to be subject to change of intercept and slope as a result of changes in technology, rate of production, or other factors which may influence change in productivity with cumulative output. Such a piecewise linear learning or experience curve can provide an instructive model of the effectiveness of technological change which permits the identification of the effect of management decision variables such as timing, scale of introduction, utilization, etc. of technological change on performance or productivity change, and the determination of their contributions. It is for this reason that an experience curve model is often used for empirical analysis of the productivity data.

2.3 MANAGEMENT OF STUDIES

The management of technological change is a relatively undeveloped area of research. Skinner [24, 25] investigated the reason for the frequent failures of departments in industrial plants to achieve organizational and planning success and identified the cause as a lack of integration and use of manufacturing policy. He concluded that for manufacturing to contribute to a firm's or industry's success, the key missions and objectives of manufacturing departments must be defined and they must be involved in the planning for technological change. Other writers, such as Wheelwright [26] and Miller [27], similarly developed conceptual models of manufacturing policy and its role in fostering changes in technology. So far, little empirical research has been performed in manufacturing policy. Buffa [28] suggested several empirical approaches.

Research dealing with the effect of stability of the environment on management decisions affecting technological change include the classic studies of Lawrence and Lorsch [9] who found that with a stable outer environment, mechanistic management systems work best, while unstable dynamic outer environments require a behavioral or organic management system. Several other studies show how environmental factors, including uncertainty, influence management decisions affecting technological change. Evidence that size and technology of the firm affect management decisions affecting technological change are suggested by the research of Dewar and Simet [29], Cummings [30], and Blau [31].

Studies using the contingency framework have shown that plants facing dynamic environments will usually show stronger preference for low levels of inventories. Also, greater co-alignment in terms of environment and policy will provide greater management decision effectiveness as shown by Skinner [24, 25].

Little research has been performed on variations in productivity among firms within the same industry. Generally the most important measures of productivity used are: labor productivity; total factor productivity (value added/total cost of labor and capital); physical partial productivity (only good for homogeneous not for heterogeneous output); and, dollar partial productivity (value output/value single output).

Measures such as partial productivity are often found to be erroneous, as productivity may depend on other factors. Partial productivity cannot usually explain differences in productivity or even measure efficiency with which a firm uses all its resources.

Fundamental problems in using productivity as a measure of effectiveness of management of technological change were found by Packer [32] to include:

a. value of counting in heterogeneous or small batch organization - subjective quality considerations;

b. focus of outputs and not outcomes similar to difference between efficiency and effectiveness;

c. trade-off of quality versus output (value of quality);

d. productivity information by its nature is inherently subjective with respect to output;

e. instead of ignoring subjective data, productivity analysts could examine their reliability using established statistical techniques from fields such as educational testing;

f. in manufacturing production rate or productivity growth is probably more important, as it provides information on trends with respect to different input factors; and,

g. measuring subjectivity is important, as the goal of productivity analysis concerns questions relating to impact on efficiency as well as creativity, flexibility, work satisfaction, etc.

There are usually problems with subjective measures in productivity analysis as it is difficult to obtain reliable responses. There is a need for some statistical analysis (bias detection, etc.) while we strive for simplicity in environment. Yet, notwithstanding these problems, productivity analysis is no less essential because it is untidy and statistically approximate. The results obviously depend on one's point of view and purpose because no single procedure will lead to objective and easily obtained numbers.

2.3.1 *Comparative Management of Technological Change.* The purpose of comparative management research is usually to develop models which could help explain the relation between different management decisions and the effectiveness of technological change in the three shipyards. The particular objective is to relate differences in management approach to decisionmaking affecting technological change and in turn the effect of management decisions on improvements in productivity achieved by technological change at the process level.

This requires a comparative study of management in different countries. Comparative management - the study of management approach on a comparative basis - is a recently developed formal discipline, though it is obviously not a new field. Early researchers in the field, such as Harbison and Myers [33], concentrated on the study of the trends of managerial development over time which could be represented by a dynamic model made up of two components of management - the people who make decisions and the framework under which decisions are made. In other words, they were concerned with the broad forces affecting the type of management in a particular environment. On the other hand, Farmer and Richman [34] developed a static model for study of management under a given set of environmental factors which produce a given level of managerial effectiveness in an aggregate economic sense. Negandhi and Prasad [35] studied management processes and their effectiveness in different environments with particular reference to developing countries with different cultural patterns, using subsidiaries of U.S. corporations as examples. Chowdhry and Pal [36] studied a textile mill in an Indian city and made some interesting discoveries of management style.

Research in comparative management deals largely with cultural and national differences, variations in communication and interpersonal relations, differences in employee attitudes and motivation, and finally differences in managerial behavior and leadership approach. Many of these factors are interdependent and raise issues of cross-cultural management comparison. England et al [37] studied cross-cultural personal values and how they affect manager behavior and in particular how their value orientation affects value judgement and management decision-making.

The effect of management thinking and resulting perception of need fulfillment, satisfaction, and importance was studied by Haire et al [38] who found that these factors influenced management decisionmaking. Bass and Burger [39] performed international comparisons of managers' objectives and impact of their life goals on their management approach.

Hofstede [40, 41] found significant cause and effect relationships between cultural environments and organizational practices in different countries, with particular reference to the influence of motivation and leadership factors. Numerous researchers have studied comparative management practices in recent years. Their findings have sometimes been contradictory. This is possibly the result of the lack of effective norms or metrices for measuring factors influencing management approaches and effectiveness.

2.4 SUMMARY

Significant research has been performed in the Management of Technological Change, particularly in the determination of the effects or impacts of technological change. Most of this work dealt with technological change at the firm or national level. Similarly, research into the social and organizational effects of technological change dealt largely with cultural or large group behavior.

In this work we carry the subject to the process and product level and the particular decision as well as individual decisionmaker concerned with effective management of technological change. We will see that there are many approaches which may offer opportunities for improvements in the decisions affecting technological change.

References

1. Mansfield, E., "The Economics of Technological Change", W. W. Norton & Co. Inc., New York, 1968.

2. Kekala, P. J., "Iron Ore = Energy, Labor and Capital Changes with Technology", J. Wiley, New York and Boston, 1973.

3. Enos, J. L., "Petroleum Progress and Profits - A History of Progress Innovation", the MIT Press, 1962.

4. Hollander, S., "The Sources of Increased Efficiency: A Study of DuPont Rayon Plant", The MIT Press, 1967.

5. Griffin, J. M., "Long Run Production Modelling - Electric Power Generation", Bell Journal of Economics, March 1982.

6. Solow, R. M., "Technical Change and Aggregate Production Function", The Review of Economics and Statistics, Vol. XXXIV, August 1957.

7. Sato, R. and Suzaw, G. S., "Research and Productivity - Prospects for the Study of Endogenous Technological Change", Auburn House Publishers, Boston, 1983.

8. Thompson, B., "Management of the Firm", Irwin, 1974.

9. Lawrence, E. and Lorsch, D., "Organization and Environment", Harvard University Press, Boston, 1967.

10. Sethi, S. P., "A Model of Management Response to Change", Econometrics, 1976.

11. Abernathy, W. J., "The Productivity Dilemma", The Johns Hopkins Press, Baltimore, 1978.

12. Wright, T. F., "Factors Affecting the Cost of Airplanes", Journal of Aeronautical Sciences, February 1935.

13. Andress, F. J., "The Learning Curve as a Production Tool", Harvard Business Review, January/February 1954.

14. Holdham, J. H., "Learning Curve - Their Applications in Industry", Production and Inventory Management, Vol. 23, November 1979.

15. Hirshmann, W., "Profit from the Learning Curve", Harvard Business Review, Vol. 42, No. 1, 1964.

16. Boston Consulting Group, "Perspectives on Experience", Boston, 1968.

17. Abernathy, W. J. and Wayne, K., "Limits of the Learning Curve", Harvard Business Review, September/October 1974.

18. Sallenave, J. P., "The Uses and Abuses of Experience Curves", Long Range Planning, Vol. 18, February 1985.

19. Kiechel, W., "The Decline of the Experience Curve", Fortune, October 1981.

20. Womer, N. K., "Learning Curves, Production Rate and Program Costs", The Institute of Management Science, November 1979.

21. Hedley, B., "A Fundamental Approach to Strategy Development", Long Range Planning, Vol. 9, December 1976.

22. Ghemawat, P., "Building Strategy on the Experience Curve", Harvard Business Review, March/April 1977.

23. Washburn, W. B., "Optimizing Production in the Presence of Learning - A Discounted Cost Approach", The Institute of Management Science, September 1978.

24. Skinner, W., "Manufacturing in the Corporate Strategy", Wiley Interscience, 1978.

25. Skinner, W., "Manufacturing Policy in the Steel Industry", Third Edition, Homewood, IL, 1979.

26. Wheelwright, S., "Forecasting Methods for Management", Wiley Interscience, 1973.

27. Miller, A. W., "Environmental Impact on Strategic Management, Addison-Wesley, 1972.

28. Buffa, E. S., "Modern Production - Operations Management", Wiley, New York, 1983.

29. Dewar, S. and Simet, D., "The Quality Circle Guide to Participation Management", Asia Productivity Center, 1982.

30. Cummings, P. W., "Open Management", AMACOM, New York, 1980.

31. Blau, A. V., "Approaches to the Study of Social Structure", Free Press, 1975.

32. Packer, M. B., "Measuring the Intangible in Productivity", Technology Review, February-March 1983.

33. Harbison, F. and Myers, C., "Management in the Industrial World: An International Study", McGraw-Hill, New York, 1959.

34. Farmer, R. N. and Richman, B. M., "Comparative Management and Economic Progress", Homewood, IL, Irwin, 1965.

35. Negandhi, A. R. and Prasad, S. B., "Comparative Management", Appleton-Century Crafts, New York, 1971.

36. Chowdhry, R. and Pal, A. K., "Production Planning and Organizational Morale", in A. H. Rubinstein, C. H. and Huberstroh, C. J. (eds.), "Some Theories of Organization", Dorsey Press, Homewood, IL, 1960.

37. England, G. W. and Lee, R., "Organizational Goals and Expected Behavior Among American, Japanese, and Korean Managers - A Comparative Study", Academy of Management Journal, December 1971.

38. Haire, M., Ghiselli, E. F. and Porter, W., "Managerial Thinking: An International Study", Wiley, New York, 1966.

39. Bass, B. and Burger, P. C., "Assessment of Managers: An International Comparison", Free Trade Press, New York, 1979.

40. Hofstede, G., "Measuring Hierarchical Power Distance in 37 Countries", Working Paper 71-32, European Institute for Advanced Studies in Management, 1981.

41. Hofstede, G., "Motivation, Leadership and Organization: Do American Theories Apply Abroad?", Organizational Dynamics, Vol. 9, 1983.

3. Economics of Production and Technological Change

Technology can be defined as knowledge used in design of products, development and use of processes, and generation of services. Economics of production is the theory used to explain the interaction of various factors such as knowledge, labor, capital, and material used in the design of products or use of processes or services in the manufacture or development of products or outputs. As shown in Figure 3.1, some of the factors used in processes or services are actually generated or exchanged by the users or consumers of the outputs of the processes of production or services. For example, consumers may offer their labor or capital as factor inputs in the production of what they themselves consume. Although the feedback characteristics represented in the diagram is a gross simplification, it does show the interaction of input factors and outputs in a consuming and factor generating environment. To represent the interaction between factor inputs used in production processes or services and the output of such activities, a production function is used.

A production function is a mathematical expression which relates the contribution of all the factors or inputs needed to achieve an output from a production process or plant to the output of such a process or plant. In other words, a production function is a statement of the interaction and interdependence of the various inputs such as labor, material, capital investments, and more needed to permit production of the desired output. For example, if labor, L, capital, K, and material, M, are all needed in equal quantities expressed in monetary value and if their contributions are additive, then the production function of an output Q (also in monetary terms) could be expressed as

$$Q = a(L + K + M)$$

Similarly if the output were equal to the product of these three inputs or factors, the production function could assume the form

$$Q = bLKM$$

Obviously Q, L, K, and M could be expressed in other terms than money, as long as these terms are consistent. In general, a production function can be expressed as:

$$Q = f(L,K,M) = \text{Output} = \text{Function of Inputs of Labor (L), Capital (K), and Material (M)}$$

for example. This usually serves as an effective model for the analysis of technology performance. To analyze the performance of a process or service, it is useful to compute the average partial productivity with respect to particular factors, say labor (L), also called the average product of labor AP_L:

$$AP_L = f(L,K,M,)/L = Q/L$$

Similarly the marginal productivity with respect to an input factor, say labor, would be expressed as MP_L where

$$MP_L = \partial f(L,K,M)/\partial L = \partial Q/\partial L$$

or the partial derivative of the output with respect to the input factor - labor.

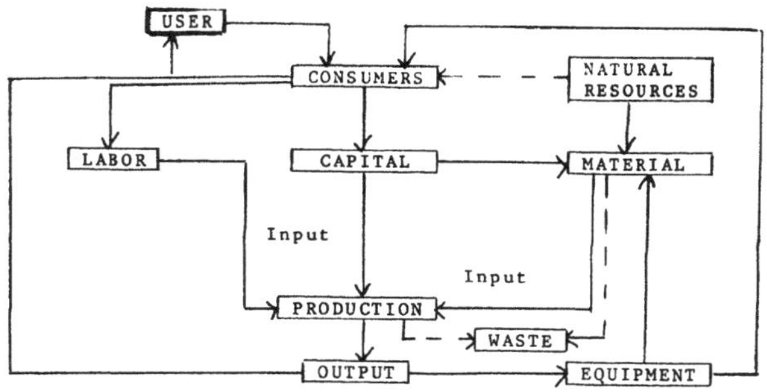

Figure 3.1. Input/Output Feedback Cycle.

The output of a production function is also called "Total Productivity"[1] (TP). The total, average, and marginal productivities with respect to the input factor L are plotted in Figure 3.2 for a simple production function to show the relationships among these three measures. Of course all these relationships depend on the values of the other production factors which are assumed to be constant while we study the above relationships as a function of the first factor.

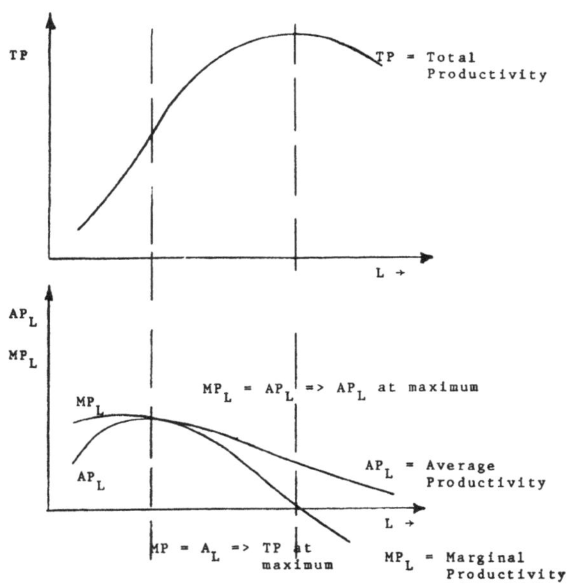

Figure 3.2. Productivities.

[1] The term "total productivity" is not used consistently in the literature. In this book it means total output as a function of all production factors.

The marginal productivity MP_L is zero at the value of the input L where total productivity TP reaches its maximum value as shown. Similarly, $AP_L = MP_L$ at the value of L at which the first derivative of AP_L with respect to L is zero, or at the value at which the AP_L reaches its maximum, as shown in Figure 3.2.

As by definition

$$MP_L = \frac{\partial TP}{\partial L}$$

$$MP_l = 0 \Rightarrow \frac{\partial TP}{\partial L} = 0$$

$$\Rightarrow \text{TP is at maximum}$$

and

$$AP_L = \frac{TP}{L}$$

$$\frac{\partial AP_L}{\partial L} = \frac{L \frac{\partial TP}{\partial L} - TP \frac{\partial L}{\partial L}}{L^2}$$

$$= \frac{L\ MP_2 - TP}{L^2}$$

$$= \frac{1}{L}(MP_L - AP_L)$$

hence

$$MP_l = AP_l \Rightarrow \frac{\partial AP_L}{\partial L} = 0$$

$$= \Rightarrow AP_l \text{ is at maximum}$$

If the marginal productivities (say in the case of $Q = f(L,K)$, MP_L and MP_K) are both positive over the range of possible values of L and K, then K and L are factor substitutes. Similarly if the partial derivative of the marginal productivity with respect to one factor, related to the other factor ($MP_L/\ K \geq 0$, or $MP_K/\ L \geq 0$) is non-negative, then the factors are called complementary factors. Finally the negative of the derivative of one factor with respect to the other (-dL/dK or -dK/dL) is equal to the ratio of the respective marginal productivities which also represents the rate of technical substitution RTS. Therefore

$$-dL/dK = (dQ/dK)/(dQ/dL) = MP_kK/MP_L = RTS$$

The RTS is therefore the amount by which K can be reduced if L is increased by one unit without changing the value of the outputs.

Another important concept is the partial or total elasticity of a production function. The partial elasticity of a production function is defined as the percentage change of the output with respect to a percentage change of an input such as labor or capital:

$$E_i = \frac{\%\ \text{Change in Total Product}}{\%\ \text{Change in input i}} = \frac{\Delta Q/Q}{\Delta X_i/X_i}$$

and similarly

$$E_L = \frac{\partial QL}{\partial LQ} = MP_L/AP_L \text{ and similarly } E_K = MP_K/AP_K$$

It is therefore simply the ratio of the marginal productivity divided by the average productivity with respect to the same input factor. This ratio also represents the percentage change in output Q obtained for a percentage change in the input factor, assuming all other input factors remain constant

$$\frac{dQ}{dL} = \frac{\partial Q/Q}{\partial L/L} = \frac{MP_L}{AP_L} = E_L$$

If all input factors change by the same percentage, then we obtain the total elasticity of the production function E which is equal to the sum of the factor elasticities or the sum of the ratios of marginal productivities divided by the average productivities for each input factor. Therefore

$$\sum_i E_i = \sum_i \frac{MP_i}{AP_i} = E = \text{Scale Coefficient}$$

which is called the scale coefficient. If the scale coefficient is unity (E=1) we say that the process exhibits constant returns to scale. Conversely if E > 1 it exhibits increasing returns to scale and if E < 1 it exhibits decreasing returns to scale. The returns to scale simply indicates if output will increase at less, the same, or larger rate than inputs.

If the production function is homogeneous, then

$$f(aL,aK) = a^E f(L,K)$$

or the output after multiplying all inputs by a constant a is a times the original output. If E = 1 we have constant returns to scale and the marginal productivity with respect to any of the inputs does not change with proportional inputs.

3.1 CHARACTERISTICS OF ISOCOST CURVES AND ISOQUANTS

Isocost curves are lines or surfaces which relate quantities of two or more factors which together have constant costs. When only two factors are used in a production process, then isocost curves are straight lines as shown in Figure 3.3. The line (or corresponding surfaces) represent all combinations of K and L which can be bought and which add to a constant cost. If the value of K in relation to L doubles, for example, then the isocost curves would be like the staggered lines in the diagram. Similarly if the value of labor L in relation to capital K doubles then the lines would be steeper. The slope of an isocost line is the ratio of the price of K to the price of L.

The output of a process represented by a production function

$$Q = f(K,L)$$

can be shown diagrammatically as in Figure 3.4. Here a combination of values of K and L can produce an output Q_i with a given technology. The curve which represents this combination of K and L which result in a constant value of Q_i is called an isoquant which has the same characteristics as an indifference curve in the consumer theory of microeconomics:

1. Isoquants are negatively curved.

2. Isoquants are convex in relation to the origin of the input variables.

3. Isoquants never cross.

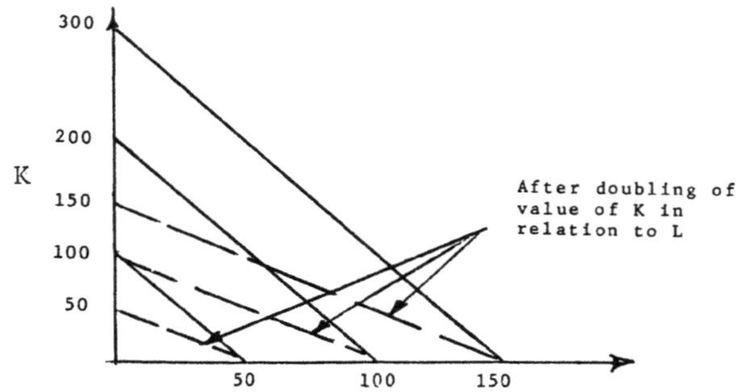

Figure 3.3 Isocost Curves.

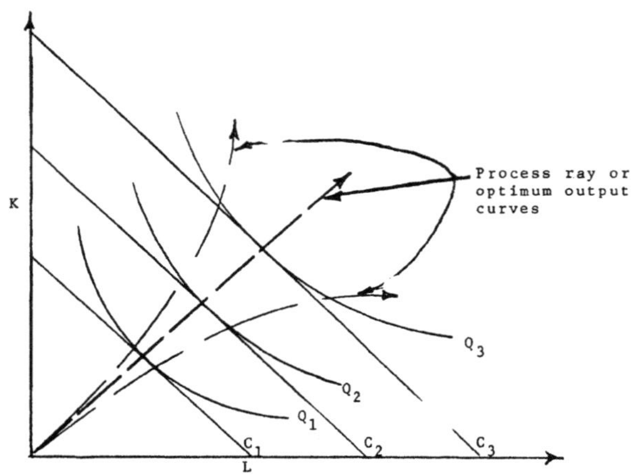

Figure 3.4. Isoquants.

The point at which isoquants tangentially touch isocost lines are the optimum point (combinations of L and K used) for the production of the respective quantity. At this point the production is in equilibrium and the output is maximized for a given outlay. At this point

$$\frac{MP_L}{P_L} = \frac{MP_K}{P_K}$$

when P_L and P_K are unit factor prices of L and K. The curve connecting these points is the optimum output curve or process ray. This curve is linear if the process exhibits constant returns to scale, i.e. is homogeneous.

If the process exhibits increasing returns to scale the curve will be concave and convex if the returns to scale are decreasing. The effect of different returns to scale can also be noted from the spacing of isoquants as shown in Figure 3.5. A process which consumes more of at least one input for the same output is considered to be less technically efficient.

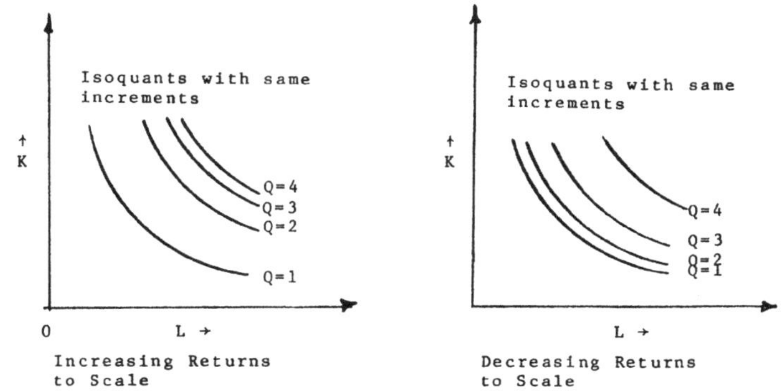

Figure 3.5 Effects of Returns to Scale on Isoquants Spacing

Different technologies (or processes) producing the same outputs will have different process rays or output curves as shown in Figure 3.6, in which three technologies, I, II, and III, are represented. Isoquants are again the curves which connect points of equal output as shown.

The isocost lines do not necessarily define a unique optimum choice of technology for all levels of output Q. The intersection may for example identify technology III as most efficient to produce Q_1 but technology II to produce Q_2 and so forth, in which case the mixed technology process line could be a curve as shown.

3.2 PRODUCTION FUNCTION THEORY

As discussed in the previous section, a production can be defined as an input/output relationship. In order to produce a commodity or provide a service, various resources are combined to yield an output. Thus, if we denote Q as the output and x_i as the ith factor input, the production function takes the following form

$$Q = f(x_1, x_2 .. x_n)$$

The inputs we are primarily interested in are capital (denoted as K) and labor (L). However, in many cases, it is necessary to consider materials (M) as well, when Q is denoted as Q = f(K,L,M).

The most known production function is the so-called Cobb-Douglas production function and has the form

$$Q = d.K^a L^b M^c$$

where

Q = the production output in $ or units of output

K = the capital used in $ or units of capital

L = the labor used in $ or units of labor

M = the materials used in $ or units of material

and

d = a constant which accounts for the agreement of the dimensions of the left and right side of the equation

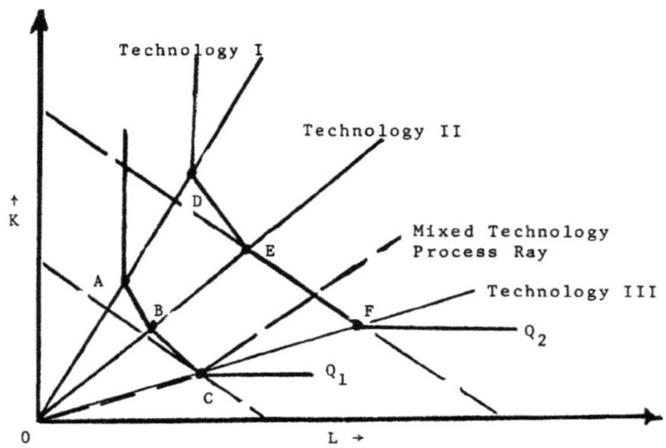

Figure 3.6 Isoquant Map for Three Technologies.

The assumption a+b+c=1 is usually made in the use of the Cobb-Douglas production function and implies constant returns to scale, as defined before.

If we assume that the price ($/unit) of each input factor is not a function of the amount of the input factor used or output produced, then, it does not make any difference if the inputs and outputs from the production process are expressed in dollars or in number of units because d remains the same independently of the volume of the entries.

As an example of the use of the Cobb-Douglas function, it is interesting to consider such a function proposed for shipyards of the form

$$V = AK^aL^b, \text{ where } a+b = 1$$

If V is the output in terms of value added, K is the capital input, and L is the labor input. To analyze the production function we fix $K = K_o$ (capital input is constant) so that we have

$$V = A\, K^a_o\, L^b$$

By letting W be the wage rate for labor and r the return of capital, we can write the total cost (materials excluded) as

$$TC = W\, L + rK$$

and solve for L to get

$$L = \left(\frac{V}{A \cdot K^a_0}\right)^{\frac{1}{b}}$$

which results in

$$TC = W\left(\frac{V}{A \cdot K^a_0}\right)^{1/b} + rK_0$$

The marginal cost can then be computed as

$$MC = \frac{\partial TC}{\partial V} = \frac{W}{bA^{1/b}\ K_0^{a/b}}\ V^{\frac{1-b}{b}}$$

and

$$\frac{MC_2}{MC_1} = \left(\frac{V_2}{V_1}\right)^{\frac{1-b}{b}}$$

where MC_1 equals the marginal cost evaluated by equation (1) when the capital is fixed at $K = K_0$, and the output is at level V_1.

For b = .796 (see Reference 1), we get

$$\frac{MC_2}{MC_1} = \left(\frac{V_2}{V_1}\right)^{.256}$$

which means that a 30% increase in output (value added) results in 7% increase in marginal costs under the assumption that the capital remains fixed ($K = K_0$). Similarly if we assume the cost of labor is fixed at $L=L_0$ then:

$$\frac{MC_2}{MC_1} = \left(\frac{V_2}{V_1}\right)^{\frac{1-a}{a}} = \left(\frac{V_2}{V_1}\right)^{2.7}$$

where a was taken to equal .271 (again from Reference 1).

In this case, however, a 30% increase in output doubles the marginal cost (see also Figure 3.7). The results, as shown in Figure 3.7, are interesting and reflect in some way the structure of the industry. From a practical point of view this result is rather intuitive. It simply says that it is cheaper and more efficient to increase the output by employing more workers and keep the invested capital constant than vice verse. Nevertheless the optimal solution lies in between: for any given level of output there is a minimum cost combination of inputs which, however, depends on the level of output (classical microeconomic theory). In the following section, we continue our discussion of production function in more theoretical terms and also try to account for the effects of technological change.

3.2.1 *Neo-classical Production Functions.* As mentioned before, a production function is a mathematical statement presenting quantitatively the purely technological relationship between the output of a process and the inputs of the factors of production. The main purpose of the production function is to display the possibilities of substitution among the inputs (factors of production) to achieve a given level of output.

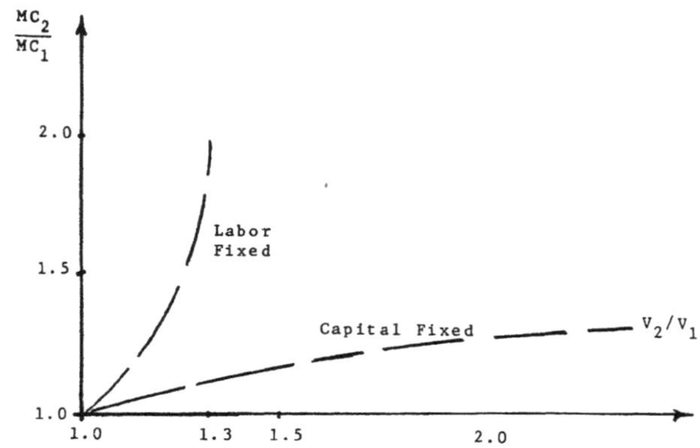

Figure 3.7. Marginal Cost in Shipbuilding.

Although most empirical work in production function theory still rests on functional forms which are the direct descendants of the work of Cobb and Douglas [2], there is an undercurrent of distrust in this theory, supported by a wealth of empirical analysis. Neo-classical estimates in contrast show that, at best, the classical theory is a gross simplification of the real world. The results of recent work on aggregation - such as those by Fisher [3] which have emphasized the stringency of the conditions that must be fulfilled before rigorous aggregation is possible, and by the growth of rival theories which have descended from the vintage models of Salter [4], all show that the traditional Cobb-Douglas type of production function is not a realistic representation of process performance.

Some particular neo-classical production functions are:

1. <u>The Constant Elasticity of Substitution (CES) Production Function</u>

$$Q = d[hK^{-W} + (1-h)L^{-W}]^{-1/W}$$

in which

- K and L are the capital and labor input respectively
- d is a scale parameter of inputs, denoting the scale on which an economy is operating
- h is the distribution parameter indicating the degree to which the production technology is capital intensive (0<h<1)
- W is the substitution parameter measuring the ease with which the technology permits labor to be

substituted for capital

2. <u>The Variable Elasticity of Substitution (VES) Production Function</u>

These are usually expressed as exponential function where d, v, and h are coefficients.

$$Q = d\ K^{v(1-hW)}\ [L+(W-1)K]^{vhW}$$

Neo-classical production functions have been widely used to explain the effect of technical progress or technological change.

By technological change, change in output can be obtained from given inputs by varying time or cumulative output when the production function can be written as:

$$Q = \phi(Y_1, Y_2, \ldots Y_m, t)$$

where t denotes an index representing the state of technology or, alternatively, time.

Neo-classical production functions can be either neutral or not. They are neutral if the balance between certain variables is undisturbed as time proceeds. They are non-neutral when the balance is disturbed. Neutrality implies that the ratio of inputs does not change with the change in output.

Various types of neutrality can be defined for a homogeneous production function in labor and capital (henceforth denoted by L and K). The output attainable from the combination of these two factors is represented by Q, while the general production function can be described as follows:

$$Q^o = F(K,L,t)$$

Considering the most important neo-classical production functions which consider technical progress:

1. <u>Hicks-neutrality (product-augmenting technical progress)</u> - Ref. [5]

Technological change is labeled as Hicks-neutral if the relationship between the marginal rate of substitution and the capital-labor ratio remains constant under time transformation. In this case, production function can be written as:

$$Q = F(K,L,t) = F[A(t)K, A(t)L] = A(t)F(K,L)$$

where A(t) represents the technical progress factor.

2. <u>Harrod-neutrality (labor-augmenting technical progress)</u> - Ref. [6]

Technological change is called Harrod-neutral if the relationship between the average product of capital and the marginal product of capital does not change. In this case, production function can be written as

$$Q = F(K,L,t) = F[K,B(t)L]$$

where

B(t) represents the technical progress factor.

3. <u>Solow-neutrality (capital-augmenting technical progress)</u> - Ref. [7]

Technological change is called Solow-neutral if the relationship between the average product of labor and the marginal product of labor remains invariant. In this case, production function can be written as

$$Q = F(K,L,t) \quad F[B(t)K,L]$$

4. <u>Factor Augmenting Technological Change</u>

Technological change is called factor-augmenting if and only if the relationship between the elasticity of substitution and the ratio of the marginal and average products of capital (labor) remains constant over time. In this case, production function can be written as

$$Q^{\circ} = F(K,L,t) = F[A(t)B(t)K, A(t)L]$$

In summary, if the production function can be expressed as

$$Q^{\circ} = F(K,L,t)$$

it can be rewritten as

$$Q^{\circ} = F(aK, bL)$$

where a and b are the capital and labor efficiency coefficients (variable with time)

In fact, augmenting technological change, three types of neutrality are defined according to how the above two coefficients change:

(a) increase in a alone	)	) capital-augmenting or) Solow-neutral
(b) increase in b alone	) technical) change is) called	) labor-augmenting or) Harrod-neutral
(c) proportionate increase in a and b	)	) product-augmenting or) Hicks-neutral

In the L,K plane, the isoquant shifts (a) vertically downward, (b) horizontally to the left, and (c) proportionately toward the origin.

3.2.2 *Vintage Model: Substitutability Before and After the Investment.* A production model with embodied technical progress is called a vintage production model. It consists of a (sequence of) heterogeneous production function(s) being constructed from the combination of inputs, which below to various "vintages". Capital vintage models, in which a vintage consists of a homogeneous set of machines produced for installation at one time and successive vintages relate to a sequence over time, have been the most commonly used to date. Often, a distinction is made between the substitutability characteristics of machines (capital vintages) and labor before and after the installation of new machines.

I. A putty-clay vintage production model is a production model with substitution possibilities before but fixed labor requirements after the machines are installed.

II. A putty-putty vintage production model is a production model where substitution possibilities between capital and labor exist both ex-ante and ex-post.

III. A clay-clay vintage production model is a production model when the capital-labor ratio is fixed both before and after capital installation.

In general, a production function expresses the relationship between various technically feasible combinations of inputs, or factors of productions, and output. It is not, however, merely a specification of the possible combinations of inputs employed in a single production process. Rather, it is a specification of all conceivable modes of production in the light of the existing technical knowledge about input-output relationship.

Figure 3.8 graphically describes technological change in the K-L, capital-labor space. If I is the original isoquant, I_1 represents the result of a labor-saving technological change; I_2 of a capital savings advance; while I_3 results from a neutral type of technological change. (This can also be defined in terms of change of MP_K versus MP_L of the different technologies.)

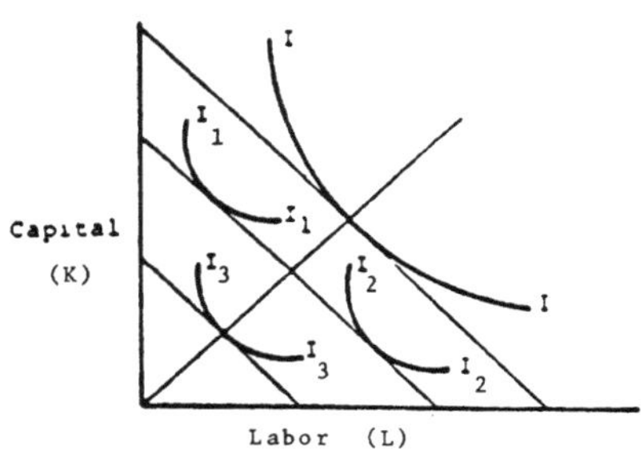

Figure 3.8. Isoquant Curves of Technological Change.

Another approach of describing technological change, proposed by Alexander et al [8] is to define a so-called technological production function. In this context, the outputs are technology and production, and the inputs are the resources or costs of development. This function can be written as:

$$D = F(T,S)$$

where

D = cost of development

T = the level of technology

and

S = shift factor that operates on the average cost curve (S is such defined that a higher value implies higher production cost)

It is assumed that there is some separability in the allocation of development resources; that is, given input can be directed in varying proportions toward increasing technology or decreasing costs.

Figure 3.9A illustrates an important assumed property of the technological production function - development costs rising rapidly as one tries to push technology beyond some limit, where Figure 3.9B shows explicitly the tradeoff between increasing technology and reducing production costs for given development expenditures.

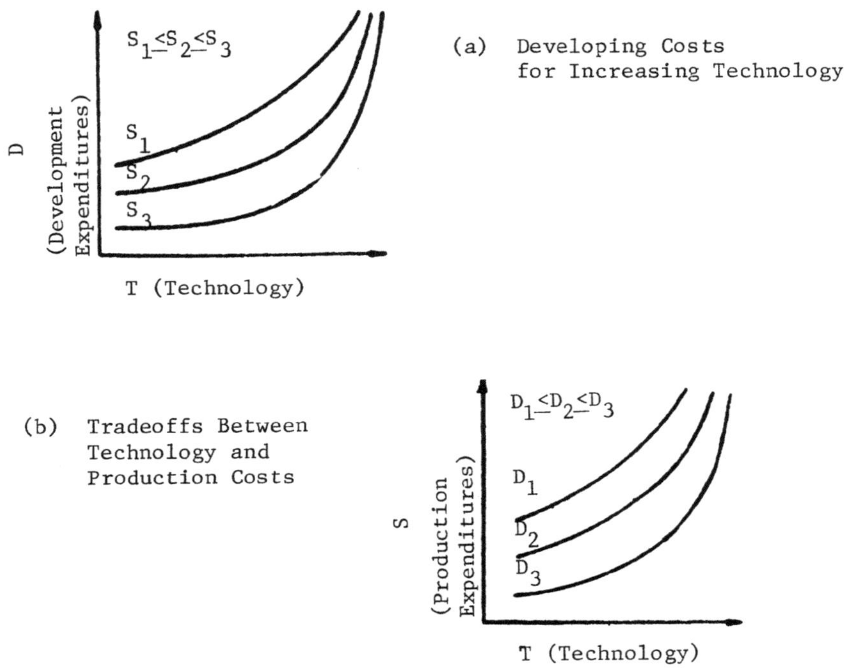

Figure 3.9. Technological Production Function, Showing Relationships Between Development Expenditures, Production Costs, and Technology. Source: Reference 2.

The technological production function moves over time as knowledge accumulates and problems are solved. This movement, therefore, depends on past development choices. An emphasis on either technology advancement or cost reduction will tend to persist because the individuals and organizations, which have become experienced at and accustomed to solving one class of problems, will find it easier to continue solving the same problems.

An example of how previous decisions may influence the movement of the technological production function is shown in Figure 3.10. Here we select one member of the family of the curves shown in Figure 3.9 and trace alternative movements of the curve over time. If technology is emphasized by a choice of Point A at time t1, the curve is likely to shift upward and to the right as shown by curves t2 and t3 (and the selected points of B and C). On the other hand, if the cost reduction is the more important goal and point 'A' is chosen, the curve may move down to the lower right, as shown by t2' and t3'.

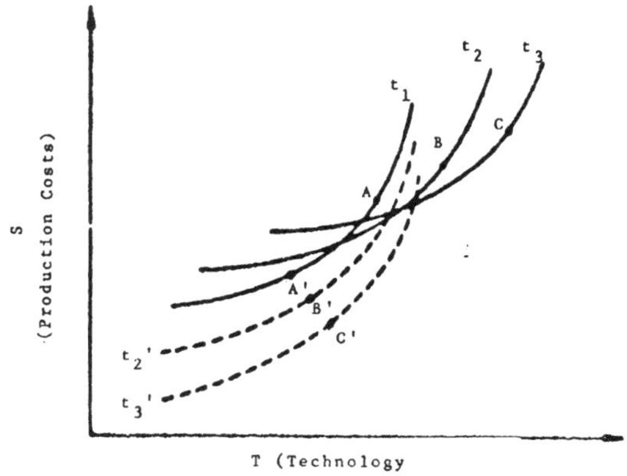

Figure 3.10. Movement of Technological Production Function Over Time as a Function of Previous Decision. Source: Reference 2.

In summary, both production function and technological production function can be used to describe the dynamic process of technological change. However, further exploration of the production function is necessary if we want to better describe and understand the production process itself. A clear understanding of the production process will enable us to identify where improvements due to technological change should be expected.

In addition to the move of technological production functions, isocost curves usually change with time, location, or political/economic developments. Such changes may result in a change of the mix of input factors in the use of a particular technology (Figure 3.11) or a change of technology used. The change of the isocost curves would result in the change in input factors from A to B as shown. Yet if another technology were available with an isoquant which touches the new isocost curves at a lower total cost such as shown as C, then a switch of the technologies and not just volume of inputs would be desirable.

3.3 COSTS OF PRODUCTION

A particular technological process is often operated at other than the most efficient conditions of input factor use. This is sometimes called operation at off-design conditions.

A production process may for example be required to employ constant labor (L) inputs over the short run, while capital (K) inputs can be changed. As shown in Figure 3.12, the process would under these conditions be able to reduce or increase output up to a maximum level, but at all except the design point, the combined input factors would not be used most efficiently, as the operating conditions would be removed from the optimum output curve.

While at the design output Q_1, the intersection of the inputs used K_2 and L* lies on the intersection of the tangent of the isoquant Q_2 and the isocost line, the intersection of inputs K_3 and L* and K_1 and L* required for output Q_3 and Q_1 respectively do not.

We consider next the multiple technology problem with say three alternative technologies capable of producing the required output (Figure 3.13). If the isocost function shown (which is the minimum cost at which Q could be produced) is the only constraint, then technology II at B provides the optimum operating condition. If in addition to the cost constraint an input factor constraint (L L*) is introduced then a mixed strategy of technology I and II at point E is the solution. When more than two constraints in terms of input factors are introduced, it is often convenient to use linear programming for the solution of the problem.

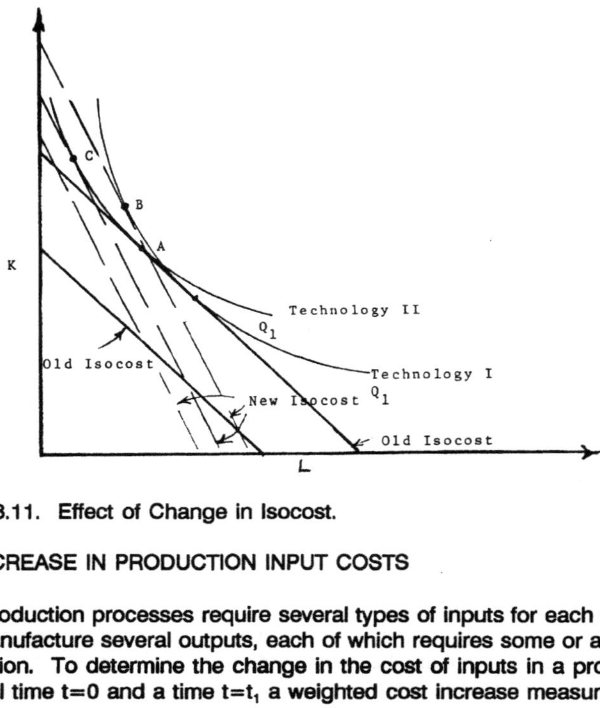

Figure 3.11. Effect of Change in Isocost.

3.4 INCREASE IN PRODUCTION INPUT COSTS

Most production processes require several types of inputs for each type of output. They may also manufacture several outputs, each of which requires some or all the different inputs in its production. To determine the change in the cost of inputs in a production process between an initial time t=0 and a time $t=t_1$ a weighted cost increase measure can be used as follows:

$$C(t_1)/C(0) = \frac{\sum_{n=1}^{\infty} I_{t_1}(n)P_{t_1}(n)}{\sum_{n=1}^{\infty} I_0(n)P_0(n)}$$

where $I_t(n)$ = quantity of input n required per unit of output at time t

$P_t(n)$ = price of a unit quantity of input n

If the process produces different outputs j (j = 1....m) where

$I_{jt}(n)$ = quantity of input n required per unit of output j at time t

$P_{jt}(n)$ = price of a unit quantity of input n

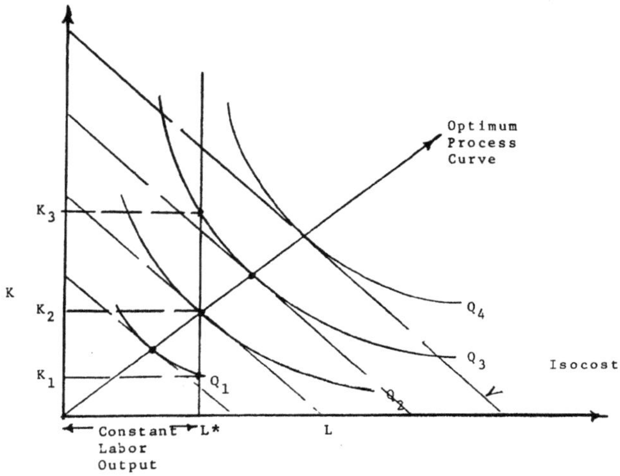

Figure 3.12. Off Design Operations.

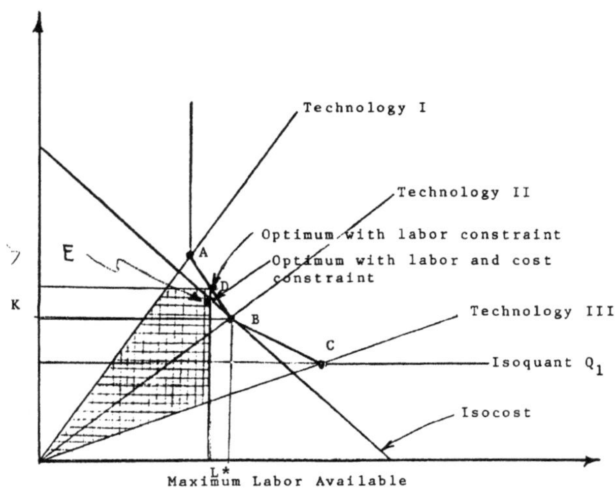

Figure 3.13. Choice Among Multiple Technologies with Cost and Input Constraints.

Then the cost increase of a unit of output type j is

$$C_j(t_1)/C_j(o) = \frac{\sum_{n=1}^{\infty} I_{jt_1}(n)\ P_{jt_1}(n)}{\sum_{n=1}^{\infty} I_{jo}(n)P_{jo}(n)}$$

and if q_{jt} = proportion of output type j at time t, then the percentage increase in costs of outputs per unit time is

$$\frac{\sum_{j=1}^{K} q_{jt}C_j(t_2)}{\sum_{j=1}^{\infty} q_{jo}C_j(o)} = \frac{\sum_{j=1}^{K}\sum_{n=1}^{\infty} q_{jt_1}I_{jt_1}(n)\ P_{jt_1}(n)}{\sum_{j=1}^{K}\sum_{n=1}^{\infty} q_{jo}I_{jo}(n)\ P_{jo}(n)}$$

where $C_j(t)$ is the cost of inputs for one unit of output type j.

The proportion of the total input costs consumed by output type j at time t is

$$\frac{\sum_{n=1}^{\infty} q_{jt_1}\ I_{jt_1}(n)\ P_{jt_1}(n)}{\sum_{j=1}^{K}\sum_{n=1}^{\infty} q_j t_1 I_{jt_1}(n)\ P_{jt_1}(n)}$$

Similar ratios can be developed to compute the unit costs of the different outputs and their percentage change in cost as well as their change in costs per unit time.

Production theory provides a meaningful framework for the analysis of the effectiveness of technology use in production of products and services. As a result, it permits evaluation of the effects of changes in technology as well as the comparative performance of alternative technologies. Production economics also permits appraisal of changes in technology in terms of their impact on society, and the environment, as well as the firm.

It is therefore a convenient analytical method for the determination of technological change decision alternatives and their expected effects.

References

1. Beazer, William et al. "U.S. Shipbuilding in the 1970's", Lexington Books, Lexington, MA, 1972.

2. Jones, H., "An Introduction to Modern Theories of Economic Growth". (A thorough discussion of Cobb-Douglas Production function is included.) Thomas Nelson and Sons, Ltd., London, 1975.

3. Fisher, M., "On the Cost of Approximate Specification in Simultaneous Equation Estimation", Econometrica 29, April 1961.

4. Salter, W.E.G., "Productivity and Technical Change", Cambridge University Press, Cambridge, 1960.

5. Hicks, J.R., "The Theory of Wages", MacMillan and Sons, London, 1963.

6. Harrod, R. F., "Economic Dynamics", MacMillan, London, 1963.

7. Solow, R. M., "Technical Change and Aggregate Production Function", The Review of Economics and Statistics, Vol. XXXIV, August 1957.

8. Alexander, C., "Notes on the Synthesis of Form", Harvard University Press, Cambridge, MA, 1964.

4. **Productivity and Technological Change**

While it is generally recognized that effective management of technological change in products and processes is critical to any firms' success and growth, and that any change in process capacity usually introduces some technological change, the management of technological change, particularly in processes, remains among the least understood aspects of operations and production management. The contradiction is that technological change is, at the same time, credited as a major contributor to the economic advance or success of firms as well as the economic growth of regions and nations.

The study of the factors which contribute to the effectiveness of management of technological change at the process level is a problem which has received scant attention. The reason is probably the lack of availability of an adequate data base of process performance under condition of technological change under different conditions of management. Similarly the theoretical basis developed for the analysis of exogenous technological change at the aggregate level, or endogenous technological change at the level of the firm, considered technological change a strategic decision concerned with the longer term impact of technology. Another reason may be the comparatively long time between technological changes at the process level in the past, which allowed consideration of process performance changes to be attributed mainly to learning. As technological change at the process level was either incremental or occurred at very long intervals, performance at the process level could not be effectively used to explain the impact of different approaches to the management of technological change on the success of the technological change, as most process technologies were replaces after long use and often after reaching obsolescence and/or the end of their operational life.

In recent years changes in process capacity or process replacement usually implied technological change. There are few processes today which do not undergo significant technological improvements even in a matter of a few years. As a result, the performance of a process over as little as 10 years will usually include technological changes introduced by process replacements or capacity additions. In fact, process technology is today subject to rapid incremental changes which occur at intervals which are small compared to the operational life of the process

This provides an opportunity for the study of the management of technological change at the process level by analyzing process performance and identifying the effect of technological change on changes in process performance. Similarly, the factors which influence the effectiveness of technological change in improving process performance should be determined and a comparative model developed for the different environments and firms studied.

4.1 EXPERIENCE CURVE ANALYSIS

Numerous studies of statistical results of productivity or production cost in relation to output from a process or plant have shown over a long time that improvements occur as a function of cumulative output. This phenomena was associated with learning and broadly discussed in the early periods after World War II. In the early 1950s it was proposed as an effective tool for the effective management of production. Its significance was highlighted though by a study of the Boston Consulting Group published in 1968, in which it was suggested that "the unit cost of a standard product declines by a constant percentage every time the cumulative output doubles". This analysis also proved that cumulative output was indeed highly correlated with experience or learning which could conveniently be expressed as a cost reduction or productivity improvement. In more recent years, the negative exponential form of learning or experience curves was established by analysis of many empirical results. At the same time, limits to the use of the experience curve were found. Among them are such obvious facts that production costs cannot decline forever and that replacement of process machinery, changes in work rules, and similar factors will often result in a change in the exponential coefficient or the slope of log of the experience curve.

Changes in process performance introduced by technological change are similarly conveniently represented by process experience curves. Statistically distinct changes in the experience or learning curves are identified by process analysis. Assume that the performance of a process i at firm j is represented by a learning curve expressed by:

$P_n^{ij} = P_1^{ij}(Q_n^{ij})^{b_{ij}}$ or in log form

$\ln P_n^{ij} = \ln P_1^{ij} - b_{ij} \ln Q_n^{ij}$

where

P_n^{ij} = productivity or cost of the nth unit made by process i at firm k

Q_n^{ij} = cumulative volume (n) produced by process i at firm j

and

b_{ij} = elasticity coefficient of process i at firm j

There are many reasons for the existence of experience or learning curves which denote improvements in process productivity or costs. These include learning, specialization, product or process improvements over time, economies of scale in production, utilization of process capacity, and rate of use of advanced methods. The slope of the experience curve of a particular process technology is usually influenced by many of these factors as well as the effectiveness of operator training, production management, and the management of the process technology. As discussed before, productivity growth in relation to cumulative output is conveniently represented by the linear relationship

$\ln P_n = \ln P_o - b \ln Q_n$

which can be used to determine the productivity (or cost) improvements as a function of the slope of the experience and the rate of growth of process output. If C_i is the cost of production of one unit when productivity is P_i then

$C_n = C_0 Q^{-b}{}_n$ = Cost of Production the nth unit

where C_0 is the cost of production of the first unit when productivity is P_0 and Q_n is the cumulative volume produced, and b is the elasticity coefficient.

If k is the slope of the log of the experience curve then

$k = C_0 Q_2^{-b} / C_0 Q_1^{-b} = Q_2^{-b} / Q_1^{-b}$ and for $Q_2 = 2Q_1$

$k = 2^{-b}$ and $\ln k = -b \ln 2$ or $b = \ln k / \ln 2$

Assuming that q_1 and q_2 are the outputs required in the first and second period respectively given Q_1 was the cumulative output at the end of the first period and therefore

$$Q_2 = q_2 + Q_1$$

then

$$C_2 = C_0(Q_1+q_2)^{-b}$$

and as

$$C_1 = C_0(Q_1)^{-b} = \text{cost of last unit produced in period 1}$$

$$C_0 = C_1/(Q_1)^{-b}$$

$$C_2 = C_1((Q_1+q_2)/Q_1)^{-b} = \text{cost of last unit produced in period 2}$$

and

$$C_{i+1} = C_0((Q_i+q_{i+1})/Q_i)^{-b}$$

If the rate of growth of output per time period is r, then

$$q_{i+1} = q_i(1+r)$$

and

$$C_{i+1} = C_0[\frac{Q_i}{Q_i} + \frac{q_i}{Q_i}(1+r)]^{-b}$$

4.1.1 *Use of Experience Curves*. Experience curves are used to study the effect of progress in a process (or product) in terms of performance which may be expressed as productivity. Productivity in such studies is usually defined in partial terms which are readily quantified by a uniform measure, such as money, manhours, etc. As noted before, the experience curve shows improvement in productivity (or costs) as a function of both slope or learning and the rate of growth of output. In other words, a technology with a lesser slope can often outperform one with a greater slope by a large rate or rate of growth of output. This fact is often used by large producers who experience competition by a small, but technologically more advanced, producer.

For example (Figure 4.1), an existing producer at point Q_E' who finds a new producer operating at point Q_N' with a cost advantage could speed up his rate of output to Q_E'' during the next period (assuming the old producer has the market), given the new producer could not exceed a cumulative output to Q_N'', at the end of the next period, to regain a cost advantage and retain market control.

A change in the slope of experience curves is often observed. This can be due to the introduction of new technology, large expansion in the use of existing technology or major improvements in the use of the existing technology by training, automation, etc. Experience curves are often used for strategic decisionmaking in such areas as:

1. investment in added capacity;
2. investment in newer technology;
3. truncating the experience curve;
4. market share and effect on strategy;
5. segmentation or concentration of process technology for mass production or to capture economies of scale, economies of experience or to assure required product line;
6. standardization - scale - focus;
7. scale of technological change; and
8. optimal time between successive expansion.

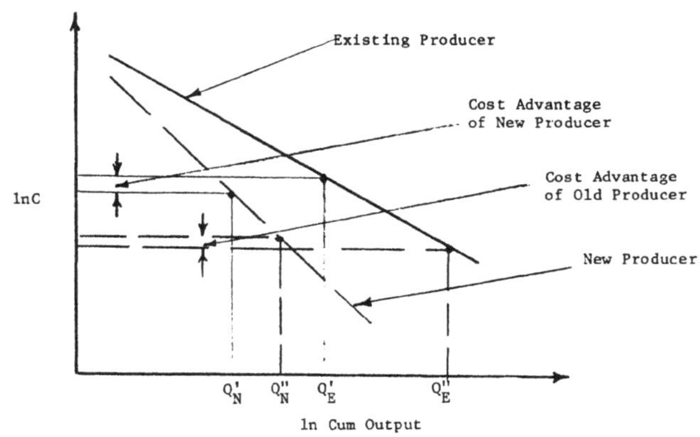

Figure 4.1. Market Control.

Decisions on expansion usually involve questions such as: when to expand? what technology to choose? lead time factors? rate of introduction? bundling rate? scale? and utilization? Another issue of interest is the learning or experience effects of quality control. Quality usually affects production cost and output. Increasing quality levels usually have correspondent effects of productivity and costs.

If there is a change in the technology or the process capacity with a resulting change in process technology productivity, then a piecewise linear experience formula can be used and

$$\ln P_n^{ij} = \ln P_1^{ij} - b_{ij}^1 \ln Q_n^{ij} \qquad n < n_1$$

and

$$\ln P_n^{ij} = \ln P_1^{ij} - b_{ij}^i \ln Q_n^{ij} - b_{ij}^2 \ln Q_n^{ij} \text{ for } n \geq n_1$$

where n_1 is the cumulative output at which a slope change occurs and b_{ij}^i or b_{ij}^2 respectively are the elasticity coefficients before and after the technology/capacity and resulting slope change.

Experience curves can be tested to determine if one (or more) distinct slope changes occur, by comparing the statistical fit of the simple and piecewise linear experience curves, and if it is shown not to be linear, computing n_1 which provides the highest improvement in the fit, if any. (In some cases, more than one slope change may be required.)

The intercepts at start P_1^{ij} and at the slope change

$$\ln P_n = \ln P_1^{ij} - b_{ij}^1 \ln Q_n^{\ ij}$$

can then be obtained, and the time t_1 equivalent to cumulative output n_1 established. The slope change $b_{ij}^1 - b_{ij}^2$ obtained for individual processes can then be used to determine the aggregate slope changes and timing of slope change for each process.

The timing of a slope change in an experience curve can usually be assumed to be a function of management decisions such as lead time (LT), bundling rate (BR), and age of new technology (T) or T = f(LT, BR, T). The timing of the slope change may be correlated with changes in input factor (such as labor) costs. Similarly the slope change S itself may be assumed to be a function of new process utilization, labor training, and capacity increase introduced by new technology.

Cross-sectional analysis of the results of the aggregate slope change and timing of slope change with respect to various possible explanatory or management decision variables can often be performed.

In cross-sectional analysis, we may use multiple regressions of the form:

$$\Delta T_{ij}(t_1) = a_0 + \sum_{r=1}^{m} a_{ri} V_{rij} \quad \text{= timing of slope change (1) for process i at firm j}$$

where

a_o = constant
a_{ri} = coefficients of decision variable r
V_{rij} = decision variables r for process i at firm j

and

$$\Delta S_{ij} = (b_{ij}^{\ 1} - b_{ij}^{\ 2}) = C_0 + \sum_{S=1}^{n} C_{Sij} V_{Sij}$$

= slope change of process i at firm j

where

C_o = constant
C_{Sij} = coefficients of decision variable S
V_{Sij} = decision variables S for process i at firm j

When new capacity is introduced into a process, the new capacity usually has a different technological level than the existing process. For example, if we add new automated welding capacity to an automated welding process it will usually include more advanced automation. As a result, the experience curve of the new capacity will have a lower intercept and different slope than the existing process. The intercept (or productivity in producing the first unit of output) of the new capacity may be lower or higher than the productivity (or cost) achieved by the existing process at the time the new process is introduced. Figure 4.2 shows the experience curves of the old and the new process. If the old process had a productivity of $P_1(Q_1)$ at the time of introduction of the new capacity (technology), and if the new capacity (technology) has an intercept productivity of $P_2(1)$ where $P_2(1) < P_1(Q_1)$ then if the experience curve of the new capacity Q_1 and the next units after Q_1 are produced by the new capacity, then the transformed learning curve of the new capacity which starts to produce Q_1 will only asymptotically converge on the original learning curve of the new capacity as shown.[1]

The transition from the old to the new capacity (technology) can take place in a number of different ways. It can be cost driven when all (or as much as possible) of the output is transferred to the new technology nearly instantaneously (Figure 4.3). Conversely, the transfer can be demand driven, when the transfer to or utilization of the new capacity (technology) is gradual (Figure 4.4).

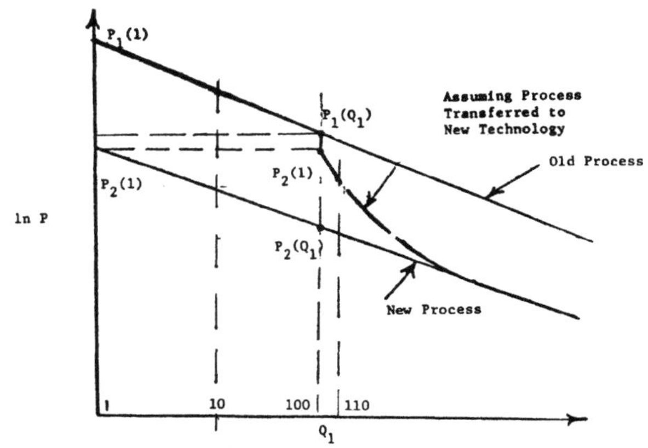

Figure 4.2. Experience Curves of Old and New Processes.

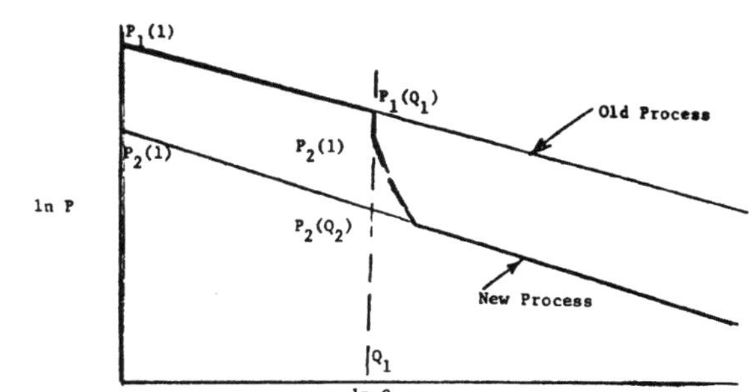

Figure 4.3. Cost Driven Transfer to New Technology.

[1] As shown in Figure 4.4, the log scale transforms the linear experience curve to an asymptotic curve if on reaching cumulative output Q, using the old process with productivity $P_1(Q_1)$ production is shifted to the new process when productivity of the next unit of output becomes $P_2(1)$ and then assymptotically approach the learning curve of the new process.

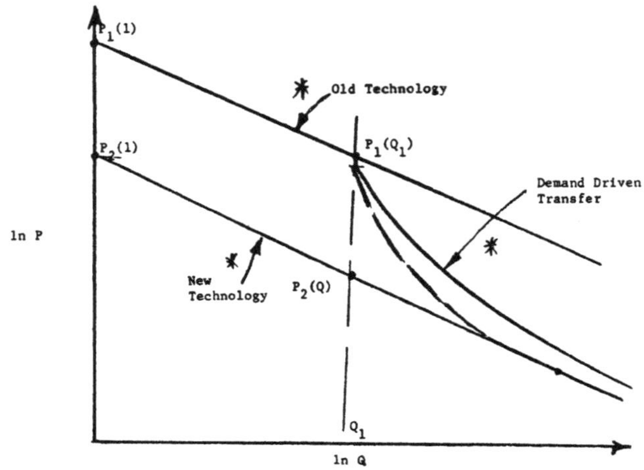

Figure 4.4. Demand Driven Transfer to New Technology.

If all the output is transferred instantaneously or gradually to the new capacity, given it is adequate, the effective experience curve will converge towards that of the new capacity (technology). If, on the other hand, only a part of the total output is transferred to the new capacity (technology), then the resulting experience curve will be a compound curve which convenes to a line between the two experience curves.

As noted, when new capacity (technology) is added to an operating process and used in the process, a transition occurs which affects the process experience curve. If the addition of new capacity and its introduction or utilization is very gradual or nearly continuous, the main effect will be a change in the slope of the process experience curve. On the other hand, if the addition of new capacity is large and instantaneous, then changes in the experience curve will depend to a large extent on the rate of introduction in use or utilization of the new capacity and its contribution to the process output. The change in the experience curve after introduction of the new technology is therefore a function of:

1. the slope and intercept of the experience curve of the new capacity (technology);
2. the rate at which the output is shifted from the old to the new capacity (technology);
3. the percentage of output produced by the old and the new capacity; and,
4. the rate of change in the output.

These factors affect both the magnitude and timing of the change in slope resulting from the introduction and use of new capacity (technology). A slow gradual shift from the old to the new capacity would create a later and smaller slope change, for example, than a rapid or instantaneous transfer of output from the old to the new process.

Assuming that Q(t) is the cumulative output of one or more process technologies producing the same type of output up to time t while $Q_1(t)$ and $Q_2(t)$ are the cumulative outputs of the old and new process technologies (1) and (2) to time t respectively, then

$$Q(t) = Q_1(t) + Q_2(t) = \sum_i Q_i(t)$$

If p(t) is the percentage of output produced by the new technology at time t or

$$p(t) = \frac{Q_2(t+dt)-Q_2(t)}{Q(t+dt)-Q(t)} = \frac{q_2(t)}{q(t)} = \frac{Q_2(t+dt)-Q_2(t)}{[Q(t+dt)-Q(t)]/dt} = \frac{Q_2(t)}{q(t)}$$

where

$q_2(t)$ = production rate by new technology at time t

and

$q(t)$ = production rate by both technologies at time t

then

$$Q_2(t) = \int_0^t q_2(\tau)d\tau \qquad Q_1(t) = \int_0^t q_1(\tau)d\tau$$

and

$$Q(t) = \int_0^t q(\tau)d\tau = \int_0^t [q_1(\tau) + q_2(\tau)]d\tau$$

If we assume productivity can be presented by an exponential function:

$P_1(t) = P_1(0)\ Q_1(t)^{-b_1}$ = productivity of old process at time t

where

$P_1(0)$ = start up productivity of old process
$Q_1(t)$ = cumulative output to time t
b_1 = slope of the experience curve of process 1

If new process capacity or technology 2 is introduced at = t, with an experience curve of the form

$$P_2(t) = P_2(0)\ Q_2(t)^{-b_2}$$

and

$$Q_1(t) = [P_1(t)/P_1(0)]^{b_1}$$

then for the combined processes

$$Q(t) = [P_1(t)/P_1(0)]^{b_1} + [P_2(t)/P_2(0)]^{b_2}$$

4.1.2 *Simple Transition Model.* A simple model of the transfer of output from the old to the new technology process can be written as follows:

$q(t)$ = output required during t to (t+dt)

$Q_1(t)$ = cumulative output of old technology to time t

$Q_2(t)$ = cumulative output of new technology to time t

$p(t)$ = percentage of output performed by new technology in interval t to (t+dt)

$P_1(Q_1(t))$ = marginal productivity of old technology at t

$P_2(Q_2(t))$ = marginal productivity of new technology at t

$P(Q(t))$ = productivity of combined process using percentage p(t) at t

also

$$P_1(Q_1(t)) = A\, Q_1(t)^{-a}$$

and

$$P_2(Q_2(t)) = B\, Q_2(t)^{-b}$$

where A and B are the intercepts and a and b are the slopes of the experience curves of the old and new process technology. Also

$$Q_1(t) = Q_1(O) + \int_0^t (1-p(\tau))q(t)d\tau$$

and

$$Q_2(t) = Q_2(T_0) + \int_{T_0}^t p(\tau)\ q(\tau)d\tau$$

where

T_o = time of introduction of new capacity (technology)

$$Q(t) = Q_1(t) = Q_2(t) - Q_2(O) = Q_1(T_0) + \int_{T_0}^t q(\tau)d\tau$$

and

$$P(Q(t)) = [1-p(t)]\ P_1(t) + p(t)\ P_2(t)$$

The relationship between P(Q(t)) and Q(t) represents an average experience curve of a mixed technology process. If production capacities are C_1 and C_2 per unit time respectively, then

$$q(t) \leq C_2 \text{ for } t \leq T_o$$

$$q(t) \leq C_1 + C_2 \text{ for } t > T$$

If $q_{(t)} \leq C$ for $T_o \leq t \leq T_1$ and a percentage of output p(t) is transferred to the new technology, then the above model can be used.

If $q(t) \geq C$, then a minimum portion of output equal to $[q_2(t)-C_2]$ must be produced by the old technology or:

$$p(t) = \begin{cases} 1 & \text{if } q(t) \leq C_2 \text{ for } T_0 \leq t \leq T_1 \\ C_2/q(t) & \text{if } q(t) \geq C_2 \text{ for } T_0 \leq t \leq T_1 1 \end{cases}$$

if $P_2(Q_2(t)) \geq P_1(Q_1(t))$, then

$$p(t) = \begin{cases} 0 & \text{if } q(t) \leq C_1 \text{ and } P_1(Q_1(t)) \leq P_2(Q_2(t)) \\ 1-C_1/q(t) & \text{if } q(t) \geq C_1 \text{ and } P_1(Q_1(t)) \leq P_2(Q_2(t)) \\ 1 & \text{if } q(t) \leq C_2 \text{ and } P_1(Q(t)) \geq P_2(Q_2(t)) \\ C_2/q(t) & \text{if } q(t) \geq C_2 \text{ and } P_1(q_1(t)) \geq P_2(Q_2(t)) \end{cases}$$

Using this model the effect of various rates of output transfer from the old to the new technology process can be studied.

For example, we can assume q(t) the output to be constant over time, linearly increasing, or sinusoidal. The model can be used to plot the resulting transition from the experience curve of the old technology for cases where the percentage performed by the new technology is:

1. p(t) = constant, which means that it instantly assumes a certain percentage of the demanded output
2. p(t) = linearly increasing from the time of introduction, and
3. p(t) = parabolically increasing from the time of introduction.

We can have any combination of changes in p(t) and q(t). For example, Figure 4.5 shows the situations where p(t) is constant and q(t) can assume any form.

If p(t) is linearly increasing and q(t) is either constant, linearly increasing, or sinusoidal, then we obtain transitions as shown in Figure 4.6. Similarly for a parabolically increasing p(t) the transitions are as shown in Figure 4.7.

In the above examples, the new technology was assumed to have equal experience in terms of cumulative output as the old technology and there is a full transfer of output from old to new technology. If the new technology is introduced at time T with no prior experience, then the old technology has already achieved experience equal to $Q_1(t_o)$. In that case for constant p(t), we obtain transition curves as shown in Figure 4.8 and with linearly increasing p(t) and constant, linearly increasing, or parabolic q(t) transition curves as shown in Figure 4.7 would be obtained.

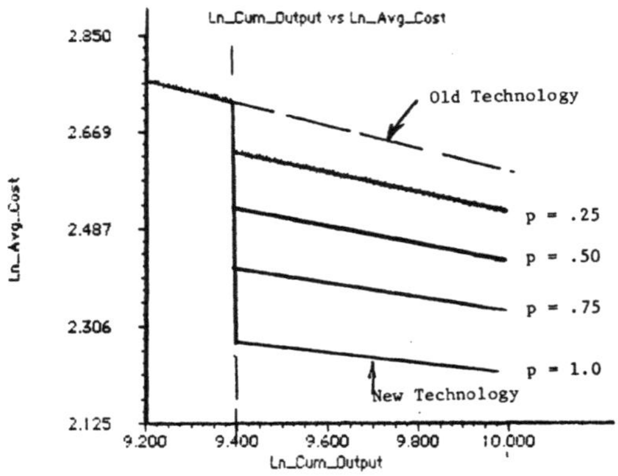

Figure 4.5. Resulting Learning Curve for p(t) constant.

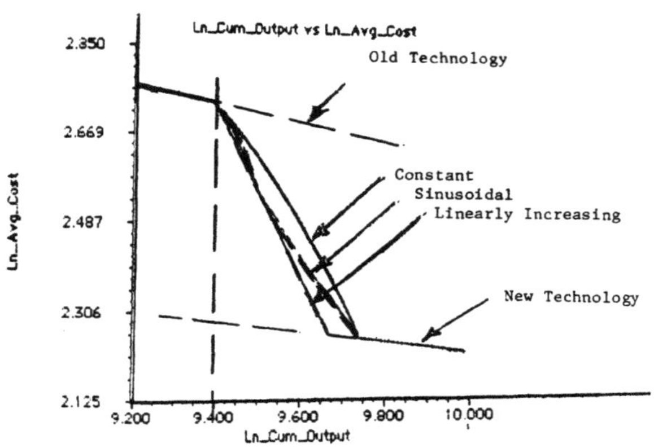

Figure 4.6. Resulting Learning Curve for p(t) Linearly Increasing.

During the transition from the old to the new technology or after the new technology capacity is introduced and available, a domain of feasible technology utilization is available to the decisionmaker. Assuming 60% of total capacity is provided by the old technology and 40% by the new technology, then different mixes of the two technologies can be utilized if output is less than 100% capacity, when each technology supplies output equivalent to its total capacity and provides 60% and 40% of the total output respectively. At lower levels of output different mixes of use of the technologies can be used. If output is less than 40% of total combined capacity, then the output can all be supplied by the new technology at the lower cost (higher productivity). Similarly output of up to 60% of total capacity could be produced by the old technology alone. If, for example, output is at 46% of combined capacity and combined unit cost (or productivity) at most equal to B is to be achieved, then the ratio of new technology to old technology contribution to that output is BC to AB. (See Figure 4.10.)

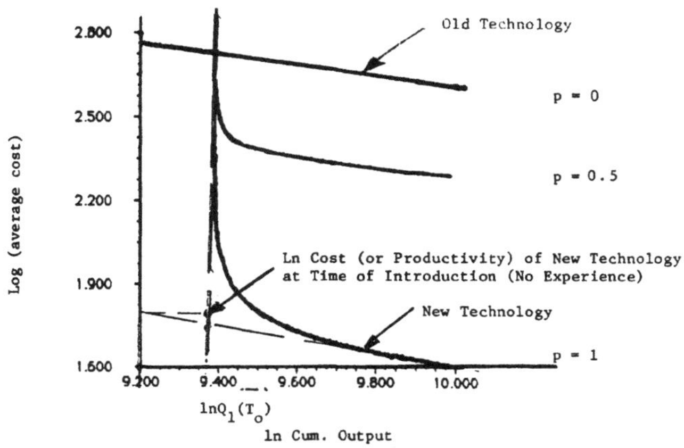

Figure 4.7. Resulting Learning Curve with p(t) Parabolically Increasing.

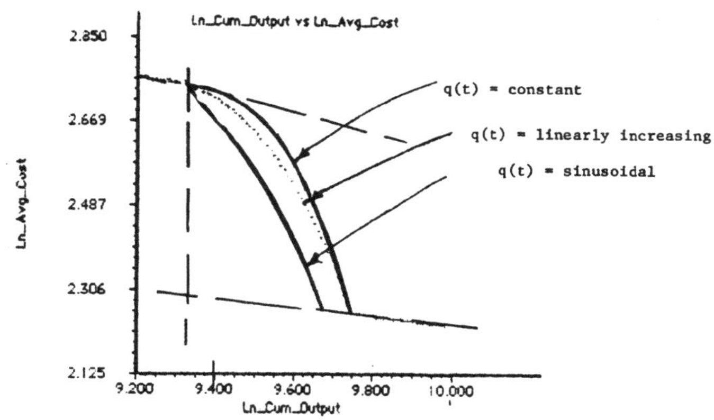

Figure 4.8. Cost Curve for Combined Use of Two Technologies - No Transfer of Experience Assumed. Percentage use of new technology [p(t)] constant.

The domain of feasible technology capacity utilization by a process which includes both old and new capacity is shown statically in Figure 4.8 for the above described example for a particular instant in time or at a particular level of cumulative output. A three-dimensional representation can be used to define the dynamic domain as affected by changing levels of utilization of old and new process technology. Obviously the rate of productivity or cost change with cumulative output (or time) will be different for the old and new process technology (experience curve slope). (See Figure 4.9.)

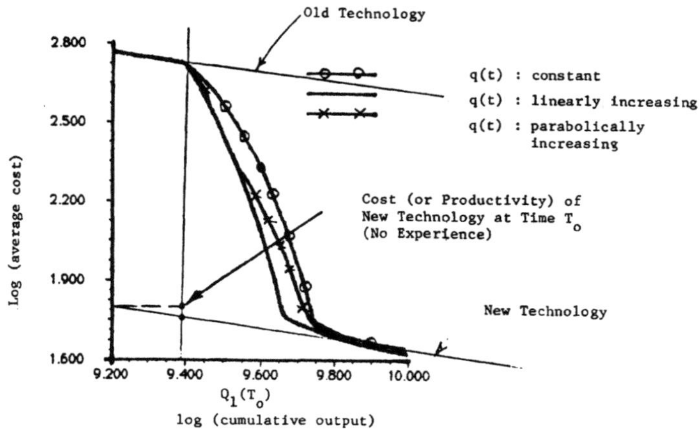

Figure 4.9. Cost Curve for Combined Use of Two Technologies - No Transfer of Experience Assumed. Percentage use of new technology [p(t)] linearly increasing.

The composite feasible domain function provides an effective means for the determination of the effects of mixed use of process technologies, and can be used to develop an optimum technology transition policy. A transition model can be built and solved by linear or non-linear optimization. This problem is non-trivial only for the case of q(t) constant, at least over the period of transition and q(t) $\leq C_1$ and q(t) $\leq C_2$. (See Figure 4.10.)

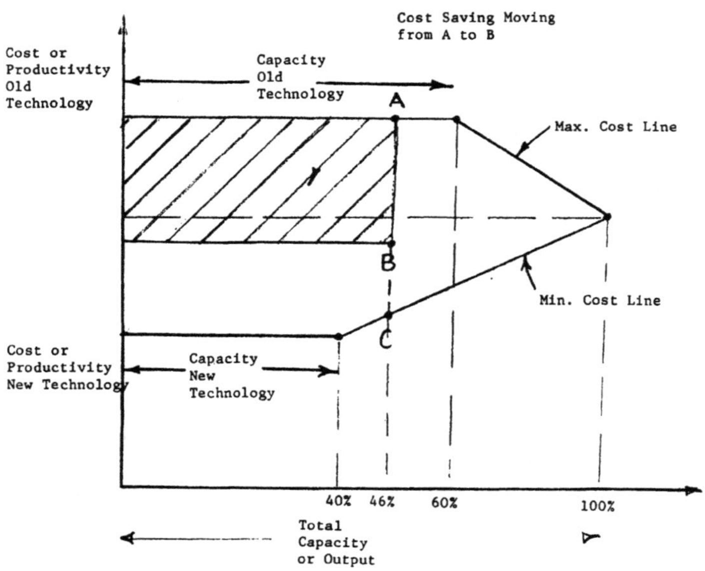

Figure 4.10. Domain of Feasible Technology Utilization.

4.2 PERFORMANCE EVALUATION AND PRODUCTIVITY MEASUREMENTS

If we are to evaluate improvements due to technological changes, we must have a datum from which to measure our progress. Therefore, we must first examine or develop measures of productivity which may serve the evaluation of the impacts of technological change. Our aim is to develop productivity measures which can be applied to manufacturing and other processes.

Productivity, by economic definition, is the ratio of output to one or more inputs which yield that output in a specific period. That is.

$$\text{Productivity} = \frac{\text{Output}}{\sum_i (\text{Inputs})}$$

It should be noted that there is a multiplicity of definitions of productivity - depending on the 'function' it is supposed to serve. Yet, almost all of these definitions can be categorized under three basic forms.

a. Partial Productivity

This is the ratio of output to one class of input, such as labor and capital, which then represents the partial labor productivity and capital productivity measures. According to recent surveys in U.S. companies, the former measure is the most commonly used partial productivity measure. However, this type of productivity measure, if emphasized alone, can be extremely misleading.

b. Total-Factor Productivity

This is the ratio of net output to the sum of associated labor, capital, and other directly employed factor inputs.

c. Total Productivity

This is the ratio of total output to the sum of all direct and indirect input factors. Since it considers the impact of all the input factors, this measure is a more accurate and realistic indicator of performance than both partial and total (direct) factor productivity.

When the productivity ratio in a current period is compared to the productivity ratio in a past or base period, we have the Productivity Index

$$I = \frac{P_t}{P_o}$$

where

P_t = productivity in year t (or current performance index)

P_o = past or base period productivity (or base performance index)

The aim of taking that further step of comparing two periods is to develop a productivity index that provides a comparative measure of performance for a product, process, or industry.

An alternative measure is the "unit manhours index", where, in the case of an industry producing a single uniform output, the Unit Manhours Index is defined as

$$I_n = \frac{1}{I_p} = \frac{1_t}{1_o}$$

where

I_p = output per manhour productivity index
1_t, 1_o = manhours/unit expended in the current and base periods respectively

In industries where there a number of products, this value becomes

$$I_{n_c} = \frac{\sum_j q_{tj} 1_{tj}}{\sum_j q_{oj} 1_{oj}}$$

also called a current period composite.

The terms q_{oj} and q_{tj} here represent base period and current output of product j, while 1_{oj} and 1_{tj} represent base period and current period labor hours/unit of output of product j.

4.2.1 *Total Productivity Index.* While this measure may provide reasonably realistic evaluation of performances, it is only logical to describe the calculation in detail. In 1972, C. E. Craig and R. C. Harris [1] presented a thesis entitled "Productivity Concepts and Measurement - A Management Viewpoint" (MIT, 1972). This work contains sections dealing with the total productivity measurement, and the four inputs considered are labor, land and capital, materials and parts, and miscellaneous goods and services. The model can, therefore, be defined as:

$$P_T = \frac{Q_T}{L+C+R+O}$$

where

P = total productivity
L = labor factor input
C = capital (and land) factor input
R = material factor input
Q = other miscellaneous factors input
Q_T = total output

For the inputs, manhours are the basic labor unit; they are multiplied by base-year wage scale values. Raw materials and purchased parts are reduced to base-year costs, estimating where necessary. Other goods and services should also be reduced to base-year values. In measuring capital as an input, a concept of "lease value" is applied, such that the capital-input factor is defined as the sum of annuity values for each asset on the basis of initial cost, productive life, and cost of capital in the base year. The fact that all costs are expressed in "base year" costs, or constant dollars, essentially creates a deflation factor for the various elements. Since inputs are assumed constant in dollar value for the current year to produce the current year volume of output, it reflects the impacts of accumulated experiences through the number of hours worked at the base year cost per hour.

On the other hand, it is essential that the organizational unit should include only the costs over which it has control. For example, in the case where raw materials are purchased at a division or corporate level, and the plant has no control over price, the Productivity Index should be calculated on a 'Value Added' basis rather than on a total cost basis.

Having described the estimation of the denominator of the productivity ratio, we must also develop some accurate measures of output in order to obtain a realistic index. Output may be defined as the sum of the products at unit selling prices and production volume for each item produced. Again, the prices should be reduced to base year levels by using base year selling price with adjustments made for new products and quality changes.

Outputs can be categorized into homogeneous and heterogeneous products. In the case of homogeneous products, the outputs can be obtained merely by counting the number of units produced. However, for heterogeneous products, we need to convert different units so that they are expressed on some common denomination. One of the weighing systems for aggregating heterogeneous outputs is to consider their required manhours for production, while another convention would be the use of value of output adjusted for price change by a price index.

4.2.2 *Computation of Productivity Change.* The next step is to calculate the percentage change (PC) in total productivity (TP) between time periods t and t-1:

$$PC_t = \frac{TP_t - TP_{t-1}}{TP_{t-1}} \cdot 100\%$$

Therefore, by assuming that there is a relationship between the change in total productivity and the usage of the productivity improvement techniques in a given period, we may be able to determine the relative impacts of the technological changes. For example, if we assume a linear relationship, where we can express PC_t as follows:

$$PC_t = A_{ot} + A_{1t}T_{1t} + \dots A_{kt}T_{kt} + \dots$$

where

$$T_{kt} = \begin{cases} 1 & \text{if technique k is used in period t} \\ 0 & \text{otherwise} \end{cases}$$

A_{kt} = productivity improvement coefficient for technique k in period t

the coefficients, A_{kt}, may be determined in the above case by Multiple Regression, using standard statistical packages like SPSS, etc.

4.2.3 *An Example of Productivity Maximization.* In this section we give a flavor of how the previous concepts can be applied. Particularly, we derive the relationship that must be reached between the amount of units of each input used, in order to maximize the total productivity of the production process.

Assume we employ the definition of total productivity as discussed before then by using letter symbols and assuming only one output (e.g. value added) we can write

$$P_r = \frac{Q}{P_K K + P_L L + P_M M}$$

where

P_r = productivity in \$ output/\$ input
Q = output in \$
K, L, M = units of capital, labor, and materials used
P_K, P_L, P_M = price/unit of capital, labor, and materials

Referring to our earlier discussion about production functions, we can express the output, Q, as

$Q = dK^a L^b M^c$ (Cobb-Douglas)

where

Q,K,L,M as above

and,
d is a normalizing and dimension adjusting constant

Now,

$$P_r = \frac{Q}{P_K K + P_L L + P_M M} = \frac{dK^a L^b M^c}{P_K K + P_L L + P_M M}$$

To maximize productivity, the following set of conditions must be satisfied:

<u>1st Order Condition</u>

All partial derivatives should equal zero or

$$\frac{\partial(P_r)}{\partial K} = \frac{\partial(P_r)}{\partial L} = \frac{\partial(P_r)}{\partial M} = 0$$

<u>2nd Order Condition</u>

The Hessian matrix should be negative definite, where Hessian matrix is

$$H = \begin{bmatrix} \frac{\partial^2 P_r}{\partial K^2} & \frac{\partial^2 P_r}{\partial K \partial L} & \frac{\partial^2 P_r}{\partial K \partial M} \\ \frac{\partial^2 P_r}{\partial L \partial K} & \frac{\partial^2 P_r}{\partial L^2} & \frac{\partial^2 P_r}{\partial L \partial M} \\ \frac{\partial^2 P_r}{\partial M \partial K} & \frac{\partial^2 P_r}{\partial M \partial L} & \frac{\partial^2 P_r}{\partial M^2} \end{bmatrix}$$

The Hessian matrix will be negative definite if the following conditions hold simultaneously:

$$\frac{\partial^2 P_r}{\partial K^2} \leq 0, \quad \frac{\partial^2 P_r}{\partial K^2} \cdot \frac{\partial^2 P_r}{\partial L^2} - \left(\frac{\partial^2 P_r}{\partial K \partial L}\right) \geq 0$$

$$\det(H) \leq 0$$

After the necessary algebraic manipulation, we obtain the set of conditions for productivity maximization, which are:

a+b+c = 1 (implying constant returns to scale)

$$\frac{M}{L} = \frac{P_L}{P_M}\frac{c}{b}$$

$$\frac{K}{L} = \frac{P_L}{P_K}\frac{a}{b}$$

4.3 DIFFICULTIES IN DEFINING AND MEASURING PRODUCTIVITY

It has become clear from the preceding discussion that although the concept of productivity appears as a simple function (ratio) of outputs and inputs it is much more complicated. There are numerous production factors that cannot be quantified as well as many aspects of outputs (e.g. quality). In the following, we list and discuss some of the difficulties involved.

a) While it will be necessary to construct production functions to estimate changes in output, great caution should be taken in establishing the capacity when using indices constructed on the basis of historical information about the relationship between capacity and capital stock. The development of more efficient production techniques, or increases in the relative price and scarcity of energy can result in accelerated obsolescence of standing capacity, but this will be missed by an index that infers capacity from capital series that are constructed on the basis of historical service lines. Likewise, pollution abatement and energy conversion investments add to the capital stock, but generally do not add to capacity.

b) The learning costs in industry need to be accounted for, and these fall into two major categories - skill and project. In the normal course of events, some skill learning is always going on in almost every organization due to labor turnover. However, project learning occurs only when a new project is started (typically a new product), or when a new process is introduced. In mass production industries, project learning is kept to a minimum by establishing highly specialized job classifications. On the other hand, longer cycle jobs, such as shipbuilding, are often so diversified that the workforce may confuse elements of different tasks which may cause interruptions to work. Therefore, one should take note of such a possibility in the evaluation of technological changes and it should be reflected in the productivity measures. This might affect the timing of technological changes.

c) The other aspect is the change in performance from experience as illustrated by the learning curve. Since the relative change in performance time between one unit to another is very significant especially in cases of small batches (or rather long cycle time lines), the phenomenon of learning curves is critical in the study of shipbuilding productivity.

d) There are other factors which will 'add value' to the products in terms of quality and attractiveness which may not be easily quantified in dollar values. However, we may model the improvements in quality as reductions of quality control costs which might be categorized into prevention costs, appraisal costs and failure costs (which reflects scrap cost, machine down-time, and customer goodwill).

e) It is unavoidable that, in a production plant, there are skilled workers and semi-skilled workers involved in the same process. In order to reflect the contribution of those skills workers, a weighing system, or conversion factor will be needed to reduce the manhours to a common denominated basis. At the same time, it must be recognized that skilled workers are more expensive. Therefore, in the case of introducing some highly sophisticated equipment, such as CAM techniques, the relative composition of the workforce might be altered, and the production function will have to be revised.

Productivity measures can be expressed in physical and value terms. In other words, we can have physical partial or total factor productivity or value partial or total factor productivity, where the latter is usually preferred because it assures use of a common measure, value, often expressed in monetary terms.

Using productivity as a measure of performance of a process or a technology introduces some possible problems or concerns:

1. Productivity measures often have a subjective quality.
2. They usually focus on outputs and not outcomes of the process studied.
3. Quality is often ignored even when productivity is measured in value terms.
4. Information on value of output is nearly always subjective.
5. Productivity measures are based on static conditions, while productivity growth is probably a more important factor because it captures the dynamics of productivity.
6. To measure productivity the approach must be consistent.
7. Productivity measurement requires the bounding of the input and output valuation or measurement to prevent double counting or omission of inputs or outputs by quantity or value.
8. The translation of different inputs into factors which can be lumped together to provide a common measure is often risky and sometimes impossible.
9. Input factors and outputs usually vary over time in terms of required factor input requirements, factor prices, etc. as must production processes and their environments are usually dynamic.

It is therefore important to develop a rational approach not only to productivity analysis but also the collection and use of labor inputs and process outputs used in measuring productivity. Most importantly, the results from such analysis must indicate the assumption made and constraints introduced by the design of the productivity analysis to prevent misinterpretation or worse misuse of the results.

While total factor productivity is theoretically a true measure of efficiency of use of resources, it is difficult to construct the divisor in a truly unbiased manner. Labor, capital, material, and other resources used may be convertible into monetary terms, but such conversion is usually only a valid index at one instant in time. Changes in factor costs, currency exchange rates, and quality or source of resource used introduces a dynamism which is difficult, if not impossible, to capture. It is for this reason that partial factor productivity measures have become popular.

Another issue clouding productivity measures is the increasing demand to include consumption of resources which do not contribute to the output (but may in fact hinder it) in the divisor of the productivity measure. These include investments in antipollution devices, social services (child care), etc. These expenditures are often loaded on the costs of individual processes, the firm, or even the industry. This brings in the question of whether a total factor productivity measure should only include input factors used directly in the process or both direct and indirect input factors. It is usually quite difficult to separate direct and indirect factors and even more difficult to properly assign and quantify indirect factors which often serve or benefit more than one process, firm, or even industry.

Very little research has been published in this area, and judgement of what to include in total factor productivity measures is usually left to the analyst who may be biased in his decision. Worse than bias may be the misuse of factors and factor valuation to achieve certain results. Currency exchange, wage cost realignments, and government aids are often used to skew results of total factor productivity analysis.

4.4 THE IMPACT OF PRODUCTIVITY

Productivity or the lack of it is often blamed for inflation, lack of economic growth, low standards of living, trade imbalances, changes in currency value, and various ills of modern society. Productivity, as a measure of effectiveness in the use of resources in production, is often misinterpreted or misunderstood. Productivity is not a theoretical concept which automatically responds to changes in resource inputs or even knowledge or technology inputs designed to improve resource use in production. It is affected not only by factors of production, including knowledge and technology (which is a form of knowledge), but also by social and cultural factors, organizational structures, regulations, environmental conditions, and individual or group incentives.

Improvements in productivity at a time of stagnant demand may cause excess capacity and as a result increased competition. Improvements in productivity should therefore be coupled with improved product quality and effectiveness so as to increase demand, which assures more effective use of the added capacity generated by the productivity improvements. While many maintain that productivity improvements should be balanced by capacity reductions, there are others who maintain that absorption capacity and thereby demand is still largely unfulfilled, and that existing demand has been stagnant because it was generated by only a small focused segment of the population towards whom products were oriented.

As products are improved with changing productivity, market scope and thereby demand can often be expanded to more than absorb the increases in output generated by productivity improvements.

References

1. Craig, C. E. and Harris, R. C., "Productivity Concepts and Measurement - A Management Viewpoint", MIT, 1972.

Bibliography

Abernathy, W. J. and Wayne, K., "Limits of the Learning Curve". Harvard Business Review, September/October 1974.

Andress, F. J., "The Learning Curve as a Production Tool". Harvard Business Review, January/-February 1954.

Boston Consulting Group, "Perspectives on Experience", Boston, 1968.

Frankel, E. G. and Chang, Y. W., "The Impact of Technological Change on the Shipbuilding Industry". MIT Report, Ocean Engineering Department 85-2, 1985.

Ghemawat, P., "Building Strategy on the Experience Curve". Harvard Business Review, March-April 1982.

Hedley, B., "A Fundamental Approach to Strategy Development". Long Range Planning, Vol. 9, December 1976.

Hirshmann, W., "Profit from the Learning Curve". Harvard Business Review, Vol. 42, No. I, 1964.

Holdham, J. H., "Learning Curves - Their Applications in Industry". Production and Inventory Management, Vol. 23, November 1979.

Kiechel, W., "The Decline of the Experience Curve". Fortune, October 1981.

Mansfield, E., "The Economics of Technological Change". W. W. Norton & Co. Inc., New York, 1968.

Sallenave, J. P., "The Uses and Abuses of Experience Curve". Long Range Planning, Vol. 18, February 1985.

Sato, R. and Suzaw, G. S., "Research and Productivity - Prospects for the Study of Endogenous Technological Change". Auburn House Publishers, Boston, 1983.

Sethi, S. P., "A Model of Management Response to Change". Econometrics, 1976.

Solow, R. M., "Technical Change and Aggregate Production Function". The Review of Economics and Statistics, Vol. XXXIV, August 1957.

Washburn, W. B., "Optimizing Production in the Presence of Learning - A Discounted Cost Approach". The Institute of Management Science, September 1978.

Womer, N. K., "Learning Curve, Production Rate and Program Costs". The Institute of Management Science, November 1979.

5. Modes of Technological Change

Technological change can be induced endogenously or exogenously or by a combination of both types of factors. Endogenous technological change is caused by actions or decisions within a system or firm, usually by rational economic decisions and attempts to allocate limited available resources under the constraints of the system or firm. If the system is a firm, then constraints on endogenous technological change may be imposed by external factors such as

1. regulatory constraints imposed by government
2. institutional constraints imposed by the company or industry structure
3. market constraints imposed by competition and market share
4. financial constraints, such as a limited credit line

and by internal factors such as

1. organization and internal management structure, or
2. technical or technological ability and capability.

In general, technological change is the result of resource allocation to an activity which may or will cause technological change. If the allocation is to Research and Development (R&D), then a risk is involved as the R&D may not produce the desired or hoped for technological change. The later the commitments of resources are made in the technology development chain: the lower the technological risk but the larger the market risks and possible costs of obsolescence. Firms will therefore assume different strategies in their resource allocation to technological change, with some investing into early stages of research and development, while others will wait and only commit resources once an invention or discovery had been made and innovated sufficiently to show or even prove its practical application and use.

Exogenous technological change is induced by external decisions, including external resource allocation as well as regulation imposed exogenously. Typical examples of exogenously induced technological change are process, product, or service developments in response to environmental laws, public concerns, or other external (often non-economic) factors, resulting in decisions affecting technological change.

Although technology is as old as the human race, it is only within the last century or two, a very short time on a historical scale, that both the rate and nature of technological change and advance has been recognized as a major factor in economic, social, and political development. More recently, it has also been recognized as a major factor influencing our environment.

Pre-industrial technology usually lacked both extensive scientific and engineering bases, and its growth was slow and incremental. It was largely developed by trial and error and it depended on muscle - human or animal - for motive power. Modern technology, on the other hand, is characterized by rapid and often dramatic growth. It often has its roots in wide and sometimes vigorous scientific research, and it applies extraordinary powerful engineering methods to its development and use.

It is clearly vital that we understand as much about the modes and methods of technological change as possible, not only for its intrinsic interest, but also for the extent to which the process of technological change can be properly controlled and guided.

An invention is usually defined as the discovery of a new product, process, material, service, or method. Inventions are the result of research and development or chance discovery. While research and development is generally planned and directed towards discovery which could become an invention, a discovery often occurs by chance and in areas not specifically addressed by the research. Similarly many inventions are the result of chance discovery not supported by research at all.

While it has been assumed for some time that there is a rigid sequence of activities leading from invention to innovation and, as a result technological progress, studies by Arrow [1] and Hollander [2] indicate that technological change is often the result of improvements in the application of known inventions. In fact many recent technological developments in production processes and products can be traced to a series of marginal improvements of the existing processes and products,including improved use and application, and often wider use or transfer of technology to areas not previously considered appropriate for application in the particular field. Research and development consists of two separate activities. Research is the process by which discoveries are attempted which would hopefully lead to inventions. Development is often defined as the process which starts to transform research discoveries into useful concepts. Development, as a result, forms the initial stage of innovation which continues as long as improvements can be made economically to the discovery. Although we generally lump research with development, many research results, including inventions discovered by research, are not subjected to development at all. In fact the vast majority of research results end up as obscure reports and papers read by a few and often narrowly defined peers who often have little interest in development, and particularly a development process which may lead to an application of the invention. This is particularly prevalent in the U.S. and Western Europe where most research is performed by academic, government, and pure research organizations. This situation has provided an opportunity for many applied researchers who are really engaged in work on development or the application of basic research results or inventions, to changes in process, product, or service technology.

For many years the path which led from invention to innovation and improvement of a technology or progress in a practical application of a product, process, or service was assumed to be the result of a sequence which often included applied research and various innovation activities. Applied research was assumed to be the principal activity which transformed an obscure research result or invention into a practical process, product, or service.

The activities from invention to innovation usually start with the identification of a potential application of the invention or research result. This is then followed by a study of the feasibility of such an application. Only then is applied research planned, after resolving any real or potential issue of rights to and conflict in the use of the invention or research result.

Innovation or applied research may take a long time and may, for example, also involve process development to permit a new product technology to be effectively employed. Innovation is therefore concerned not only with the development of a practical, marketable application of the invention or research result, but also with the production, design, distribution, and marketing process. In other words, it includes engineering and design, product or process specification and development, test production process establishment, and actual trial manufacture of the product or physical test of the process.

The time from identification of a discovery or invention as being potentially commercially useful to the introduction of a resulting innovation in a commercial form may take as little as 2-4 and as much as 8-15 years now. Much depends on the approach taken, the support available, and the degree of proliferation or separation of all the related activities leading to innovation. The time expended is often a function of market opportunities and therefore market-pull factors [3, 4]. The pressure exerted by competition may induce accelerated efforts to develop new technology which offers opportunities for cost reduction or other attractive features. Large firms or government agencies usually require more time to bring innovation to fruition than small firms, working with diverse outside groups and experts. These are often able to use more flexible and efficient approaches in the process of innovation.

Once innovation is completed because no further improvements can be developed economically or has advanced sufficiently, diffusion of the product or process involved often becomes general. The rate and timing of diffusion of an innovation depends on many factors:

a. effectiveness of the innovation in improving productivity of a process innovation and increasing market share of a product innovation

b. profit potential introduced by the innovation in terms of potential cost and price changes offered by the innovation

c. the size and composition of the industry in which the technology is used and competes

d. the degree of competition among industry firms

e. technology communication among industry members

f. ability of managers to interpret implication of innovation

g. constraints on effective use of innovation such as work rules, government regulation, market prejudice, etc.

Diffusion of an innovation can similarly start at any time during the innovation process and its development, scale, and rate will largely depend on: ownership of innovation; usefulness of innovation; cost of innovation; knowledge of innovation; and market and competitive aspects.

It is clearly vital that we understand as much about technological change as possible, not only for its intrinsic interest, but also to learn how it can be properly planned, controlled, and guided - in terms of the timing, rate of technological change, and control of the forces that impel those changes. In recent years, the management of technological change has become the single most important function and responsibility of decision makers in government, industry, and the public at large.

The rate and direction of technological change are heavily influenced by the following sets of factors.

1. Demand for a product, service, process, or method: the promise or prospect of growing demand which may attract inventive or innovative activity, and the investment required to sustain it.

2. Cost of existing technology in terms of factor inputs: technological change tends to focus in such a way that it usually affects those factors which are major cost elements or whose costs show the sharpest increase, and thereby make the affected technology less and less attractive.

3. Competition: service, process or product competition in the market often tends to provoke a defensive and innovative response which often leads to technological change.

Competition can be personal, local, or even national or worldwide, and may be similar or drastically different technologies.

4. Technological competence: technological competence of a firm, group, or individual or suppliers affects the rate of technological change, and as a result market change and competitiveness.

5. Endogenous harmony and consistent cooperative environment.

6. Environmental and exogenous support of technological change.

The pioneering work in the development of both qualitative and a quantitative analysis of technological change and its impact on growth and development was performed by Mansfield [5] who examined the time it took for a new technological development to spread from company to company within particular industries such as iron and steel, brewing, rail transport, and coal mining, as discussed in section 5.4, and also identified the different modes or stages of the technology life cycle.

As shown in Figure 5.1 technology grows from initial development which includes the R&D stage and initial innovation to technology application which is usually achieved in the later phase of innovation. Thereafter technology application is launched followed by application growth, technology maturity when few, if any, improvements or new penetrations (applications) occur. Finally, maturity results in a downturn and degradation and ultimately abandonment of the technology.

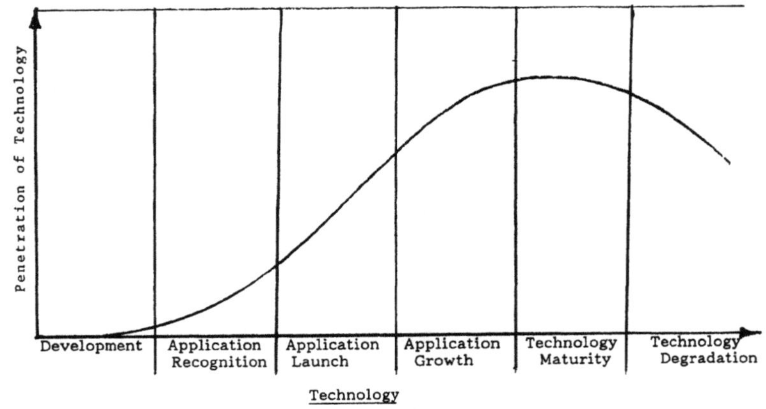

Figure 5.1. Technology in Life cycle.

Technology development consists of research based on recognition of potentially valuable technology or discovery with a consequent identification of potential uses of technology based on discovery. Development usually continues by innovation if

1. there is an obvious application and/or identifiable market;
2. resources are available for the development;
3. there are potential sponsors, supporters, or buyers of the innovation; and,
4. the risk in the technology of development is low and the expected rate of return is high.

At this, as well as at later stages, an important question is always whether to develop the technology alone, find a partner in the innovation or development, or sell it to a future user at an appropriate stage in the innovation process.

Technology application may start early during the innovation process if (1) the potentials of the applications for competing technology developments are high, (2) the costs of such application are reasonable, and (3) a market can be identified at this stage. There are usually no easy answers if and when to start the application of a new technology, during the process of innovation. Important considerations involve:

1. incurred and additional innovation costs expected;
2. cash and credit positions;
3. relative priority and alternative use of resources;
4. potential market for technology; and,
5. essentiality of technology.

Finally, the decision also involves sale, license, or in-house developments and methods used for the launch of applications of the new technology. The key issue in the innovation to application process is timing. When a technology has been diffused and has been in general use it may achieve maturity and very little further investment is justified. The principal objectives at that stage are often to delay or defer use of new technology so as it extend application and use of an existing technology as far as possible with little, if any, further resource commitments.

All stages in the life cycle of technology are subjected to endogenous and exogenous factors which influence the decisions affecting the stage and ultimately the development and application of the technology. These factors have caused a substantial delay in the development of some essential technologies, while others have been buried before effective development only to be occasionally rediscovered long after.

5.1 ENDOGENOUS FACTORS AND THEIR EFFECT ON TECHNOLOGICAL CHANGE

With the exception of the lone inventor or chance discoverer, most technology development takes place inside an organization with its own structure, culture, history, capacity and objectives. Organizations also often assume roles and try to perpetuate a perceived function. All technology development, be it by individual inventors or discoverers or in an organization, is subjected to endogenous factors which influence the behavior and decisionmaking approach of the individual or the organization.

Individuals are affected by their background, surroundings, financial situation, cultural upbringing, peers, and family support, and more. Similarly organizations are influenced by endogenous factors such as:

1. organization
2. financial condition
3. worker/management relations and worker attitude
4. worker incentives
5. worker training
6. management approach to decisionmaking
7. company culture
8. process condition

Other important endogenous factors are an organization's history, the leadership of its management at all levels of management, interpersonal relationships, work environment and quality, the prestige generated by the firm, and the effectiveness of endogenous communications.

The role that technological change is perceived to play, not only by management but also all workers of a firm, is extremely important. If everyone knows that comments, ideas, even criticism will not only be taken seriously and encouraged but also be acted upon without any possible retribution and with a potential reward if it results in an improvement, then and only then will every member of a firm pull together and share experience, knowledge, and ideas for the firm's good.

Technological change in a firm, to be effective, requires close cooperation among staff in the discovery of a need for change and a technological approach for the resolution or solution of the need, problem, or opportunity. Such cooperation is even more essential in facilitating the process of innovation and application of new technology.

5.2 EXOGENOUS FACTORS AND THEIR EFFECT ON TECHNOLOGICAL CHANGE

Technological change is affected and often even induced by exogenous factors, most of which are outside the control of the technological change decision makers. They may consist of:

1. institutional structures or organizations;
2. regulatory and other constraints or incentives;
3. cultural contexts;
4. government policy and role (type of government);
5. government ownership and control (obtrusive or non obtrusive);
6. political climate including political risk;
7. labor role, relations, and organization;
8. industrial or global position;
9. ecology;
10. prestige and expectations; and,
11. technology transfer controls.

Exogenous factors are usually pervasive and often introduce major obstacles to effective technological innovation and application. While many exogenous factors are designed to protect the public by preventing abuse, safety hazards, noxious, and otherwise undesirable characteristics of processes, services, or products offered by a new technology, some factors may be used to delay technological progress.

While it is true that not all technological change leads to or even constitutes progress, very little economic and technical advance has been achieved without technological change. Also the expectation that technological change will result in economic improvements, depends to a large extent on the assumption that variables such as price, market demand, and competitive forces remain constant - which they seldom do.

Exogenous factors are not only real but must be recognized as the principal factors which account for the difference between the expected cost savings in input factors and other costs (in a process for example) as well as benefits from improved quality, design, or performance and the economic success actually achieved by a technological change. In other words, exogenous factors account for most of the difference between expected technical and achieved economic performance of a technological change.

The elasticity of the market, for example, is an exogenous factor which will influence the economic success of a technological change, particularly if the change not only affects costs but is, as a result of expected performance, used to increase scale and thereby market share. In fact, market share may not increase even with larger scale, lower cost output if demand is highly elastic.

Tightening of environmental requirements, on the other hand, is an exogenous factor which often not only reduces but even negates cost savings and other improvements offered by a technological change. It is therefore important to consider both technical and endogenous, as well as exogenous, factors in evaluating the economic advance offered by a new technology.

5.3 MODES OF TECHNOLOGICAL CHANGE

There are many different modes of technological change and their development may be affected by the various endogenous and exogenous factors discussed. Technological change may occur by restricted technology developer or owner application without diffusion to or adoption by others. The use in fact may be strictly controlled for much, if not all, of the life of the technology. This is particularly prevalent today, when many technologies have a significantly shorter life than the life of patent protection in the U.S. for example which extends over 17 years. This was not always so, and the duration of patent protection was actually designed to permit the inventor and/or innovator technology developer sufficient time to recoup the investment in technology development and then diffuse the technology in a broad field of adopters.

While many developer/owner applications substitute new technology for old and the purpose of the new technology development is to find an improved substitute for an older process or product, there are also many who use new technology for new, often original, purposes and not substitution for an existing technology. Similarly adoption of a new technology by others not involved in its development may be by licensing or other method of transfer or diffusion. Again the new technology can be used to substitute for an existing technology or for new purposes which often involved development of a new market.

If the new technology replaces existing technology then the rate of diffusion will affect the rate of substitution. If diffusion is constrained by technology protection, then the rate of substitution by the technology developer/owner will depend on his position, objectives, market share, and various other factors not dissimilar from those of a firm who considers substitution as part of a diffusion or adoption process. In general, the principal factors affecting the rate at which new technology replaces existing technology depends on

1. investment requirements
2. advantages and profitability improvements offered
3. market conditions (breadth, competitiveness, dynamics, etc.)
4. market position
5. time since introduction of existing technology
6. knowledge of new technology available
7. conception of risk in adoption of new technology
8. environment and exogenous factors
9. acceptability by labor, organization, and endogenous factors.

Another important consideration is the expected relative performance and range of performance of the new versus the existing technology. New technology often not only offers a short and long run improvement in cost or performance over that of the old technology, but also may be able to perform or serve in ways the existing technology was never able to.

5.4 DIFFUSION AND SUBSTITUTION OF TECHNOLOGY

Technological diffusion, as mentioned, is the process by which use of application of a technology is broadened. Mansfield [6] did the pioneering work in the quantitative analysis of diffusion and substitution of technological change. For each of four industries - bituminous coal, iron and steel, brewing, and railroad - he examined the time it took for a new technological development to spread from company to company within a particular industry. His results indicated that after a new technology was adopted by one or two companies, it spread to other companies at an initially slow rate. After that, this rate increased rapidly for a time, then slowed down once again, levelling off as all the companies would adopt the new technology did so.

Studies of this type have been performed for a number of new technologies, ranging from catalytic cracking processes to numerically-controlled machine cutting, and also for the international diffusion of technology as a whole. The results are similar for most industries. Plotting the number of firms who have adopted a new technology as a function of time, we obtain an S-shaped curve, as shown in Figure 5.2, which shows the variation in the rate of adoption or diffusion of new technology. It may be useful to make some comments on technological diffusion.

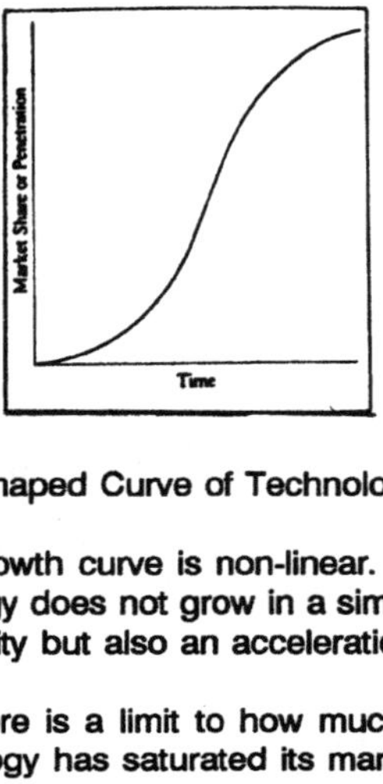

Figure 5.2. S-Shaped Curve of Technological Diffusion.

First, the growth curve is non-linear. The diffusion or use and resulting market share of a new technology does not grow in a simple progressive way. The technological change has not only a velocity but also an acceleration.

Second, there is a limit to how much a new technology can spread. It is obvious that when a technology has saturated its market, it cannot grow any further.

While the S-shaped curve gives a good general picture of how technology grows, the specific values of the takeover time (the time it takes a new technology to move from a 10 to 90 percent share of the market) are determined by the forces behind the technology. Some work has been done on the effects of such factors as the rate of return on investment, the cost of introducing a new technology, and general economic conditions.

It is important to remember that, under normal circumstances, when a new technology gains, an old technology loses. The S-shaped curve is the result of new technology growing at the expense of the old one and, indeed, often at the expense of other competing new technologies as well. Therefore, the consequence is, when a new technology comes along, some firms are able to take advantage of it while others are not. The reward for the winner is great, the penalty for the loser is often severe.

It is also important to recognize that technology advances are usually possibie only if major investments are made in research and development. In other words, for technological change to occur, scientific need discovery and engineering method must be nurtured. Each new generation of technology or new kind of product or production requires more research, experimental development, motive processes, and process and engineering developments.

A strategic relation expressing these dependencies was developed by Dobrov [6] as follows:

$$\frac{dS}{Sdt} > \frac{dT}{Tdt} > \frac{Dp}{Pdt}$$

where S is science, T is technology, and P is production.

The following is an example which can help us to understand the foregoing relationship among science, technology, and production (Figure 5.3).

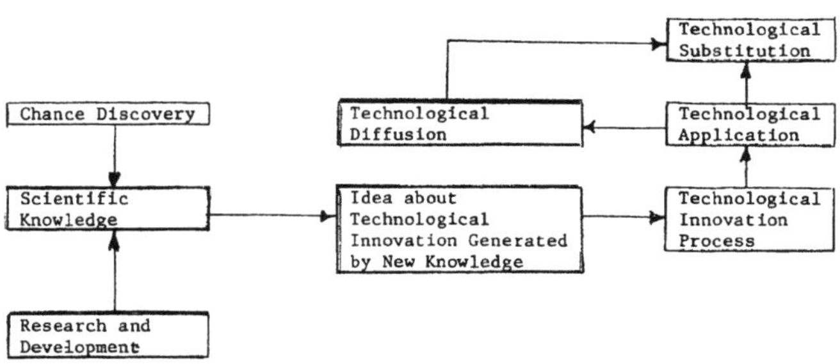

Figure 5.3. The Flow of Ideas from Science to Technology to Production.

During the 1930s, scientists with a knowledge of the latest development in atomic physics, predicted that this knowledge could be used to produce energy in the future. In 1924, the existence of the neutron was predicted by Rutherford and Chadwick on theoretical ground, and in 1932, it was discovered experimentally by Chadwick. It was soon realized that neutrons could be used to induce artificial fission and chain reactions. It was also realized that if the yield of neutrons could be controlled, the chain reactions could also be controlled. This idea opened the way for the development of nuclear reactors. This example provides a general scheme for the flow of ideas from science to technology and to production as shown in Figure 5.3.

This example also shows how the growth of knowledge stimulates the technological development, therefore increasing accumulation of scientific development tends to accelerate the technology substitution rate, as shown in Figure 5.4.

By using investment in scientific development as the input and increase in productivity as the output, Dobrov [6] proposed the following formula to calculate B, the benefit of technological change.

$$\frac{E(i)T(i)N(a)}{Q(r)T(r)K+Q(t)T(t)N(t)} \tag{5.1}$$

where

E(i) is the effect of using the ith new technological system for one year

T(i) is the time for using this technology before substitution

N(a) is the average number of technological systems in operation during the time T(i)

Q(r) is the average cost of one year of R&D for preparing this new technology

T(r) is the time of R&D needed for preparing this new technology

K is the coefficient of multiplication and proportion of unsuccessful R&D projects

Q(t) is the average cost of one year of the process of plant and market introduction of this new technology

T(t) is the time needed for introducing one technological system

and

N(t) is the number of introductions needed for transferring this new technology.

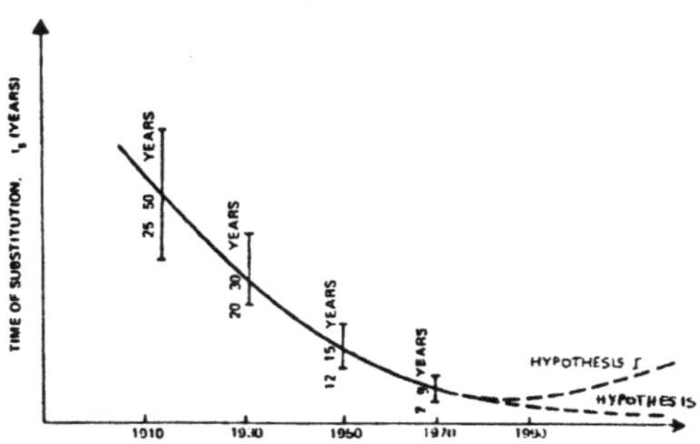

Figure 5.4. Trends in Technology Substitution Time

Dobrov's benefit equation, shown above, relies on two classes of technological indicators. One of them concerns the inputs to R&D, for example, scientific and engineering manpower and other R&D expenditures; the other one concerns the outputs of R&D, for example, publications, citations, and patent counts. Unfortunately, in practice, difficulties arise in the interpretation of those inputs to R&D, such as scientific and engineering manpower. It is not merely that scientists and engineers can vary greatly in terms of the quality of their work. The measurements of R&D outputs suffer from drawbacks of their own. Consider, for example, patent statistics. As is well known, a great number of patented inventions do not result in innovations and subsequent use and diffusion. Moreover, not all inventions are patentable. Counts of scientific and technical publications suffer from many limitations of the same sort. However, Equation 5.1 does reveal that R&D activities are vital to technological advance.

5.4.1 *Substitution Models for Technological Change*. After a new technology is introduced by one firm, other firms in related fields will usually follow. This process can be slow or fast and involve many different considerations. Many researchers have investigated the factors which determine the rate of substitution of new technology. Substitution here implies replacement of an existing technology by the new technology. In most cases, new technology does not exactly replace existing technology, as it often changes the scope, quality, and market of the new technology. Mansfield [5] developed a fundamental model of technology substitution which studies the problem from the point of view of market share.

If we assume m(t) to be the market share captured by the innovation at time t, and if L is the upper limit of the market share which the new technology can be expected to capture in the long run, and P(t) is the portion of the market captured by the new technology between times t and t+1, the following relationship obviously holds:

$$\lambda(t) = \frac{m(t+1) - m(t)}{L - m(t)} \tag{5.2}$$

It is assumed that $\lambda(t)$ is a function of (1) the proportion of market where the technology is already introduced by time t, (2) the expected profitability of the new technology P(t) (3) the amount of the required investment, and (4) other factors A. So, we can generally write:

$$\lambda(t) = f\left(\frac{m(t)}{L}, P(t), I(t), A\right) \tag{5.3}$$

This function would normally be affected by the behavior of the following factors.

1. As the proportion of market captured by the new technology m(t)/L. As this ratio increases, subsequent increases will accelerate because as more information and experience becomes available the risk of substitution of the new for the old technology becomes less and less.

2. The profitability (P(t)) of the new technology should also influence λ (t). The more profitable the investment (the greater the P(t)), the higher the probability that the use of the new technology will be extended.

3. For equally profitable innovations, $\lambda(t)$ should tend to be smaller for those innovations requiring relatively larger initial investments.

4. For equally profitable new technologies with the same investment requirement, $\lambda(t)$ is likely to vary among different industries. The value of λ (t) may depend on the degree the particular industry or firm in that industry is risk-taking or not, on the competition within the industry, on the attitude of the labor force toward the new technology or on the financial health

of the industry. Generally, it has been observed that inter-industry differences may have a significant effect on $\lambda(t)$.

Assume, as would normally be the case, that within the ranges of interest, $\lambda(t)$ can be adequately approximated by a Taylor's expansion that drops third and higher order terms. Assuming that the coefficient of higher order terms such as $(m(t)/L)^2$ in this expansion is zero (the actual data supports this assumption [6]), we obtain:

$$\lambda(t) = a_1 + a_2 \frac{m(t)}{L} + a_3 \cdot P(t) + a_4 \cdot I(t) + a_5 \frac{m(t)}{L} + a_6\, I(t) \frac{m(t)}{L} + a_7 \cdot P(t) \cdot I(t) \ldots$$

and

$$m(t+1) - m(t) = \{L - m(t)\}\ (a_1 + a_2 \frac{m(t)}{L} + \ldots. + \ldots.)$$

Assuming now that time is measured by fairly small units, we can use, as an approximation, the corresponding differential equation

$$\frac{dm(t)}{dt} = [L - m(t)]\ \{Q + \phi \cdot \frac{m(t)}{L}\}$$

the solution of which is:

$$m(t) = L \frac{\{e^{1+(Q+\phi)t} - Q/\phi\}}{1 + e^{1+(Q+\phi)t}} \qquad (5.4)$$

where L is a constant of integration, Q is the sum of all terms in the Taylor expansion not containing m(t)/L, and

$$\phi = a_2 + a_4\, P(t) + a_6\, I(t) + \ldots \qquad (5.5)$$

is the coefficient of m(t)/L.

One simple condition that we can impose on the above solution is to require that

lim m(t) = 0

From this condition, it follows that

$$m(t) = L\{1 + e^{-(1+\phi t)^{-1}}\}$$

Thus the growth of the market share over time captured by the new technology conforms to a logistic function, an S-shaped growth curve. Therefore, if the assumptions we have made so far are correct, it has been shown that the rate of introduction of new technology is governed by only one parameter ( ). In addition, assuming that the sum of the unspecified terms is uncorrelated with P(t) and I(t) and that it can be treated as a random error term, we can write

$$\phi(t) = Z + a_4 \cdot P(t) + a_6\, I(t) + \text{error (E)}$$

where Z equals a_2 plus the expected value of the sum of unspecified terms and E is a random variable with zero expected value. So, the expected value of ϕ in a particular industry is a linear function of P(t) and I(t)

$$\phi(t) = Z + a_4 P(t) + a_6 I(t) \tag{5.6}$$

Mansfield [5] checked this model using data on the substitution of 12 new technologies in four industries (brewery, coal, steel, and railroads) within the period 1900-1960 and he obtained good results. Moreover, assuming that a_4 and a_6 do not vary among industries, he found that

$$\phi = Z + 0.53\ P(t) - .027\ I(t)$$

with a very high correlation coefficient.

5.4.1.1 The Innovation Index

An index which will provide an indication of the innovation characteristics in various industries can be developed. Such as index is designed to be a single measure of the innovation characteristics of an industry and it is obtained by aggregating statistics related to the innovation process into a meaningful summary statistic.

Common factor analysis can be used to achieve the aggregation of the data into a summary statistic.

If we use factor analysis, each of n observed variables is described linearly in terms of m new uncorrelated common factors:

$$F_1\ ,\ F_2\ \ldots\ F_m \text{ and unique factors } U_j,\ (j = 1,2,\ldots n)$$

$$\begin{aligned} Y_1 &= a_{11}F_1 + \ldots\ldots + a_{1m}F_m + b_1U_1 \\ &\vdots \\ Y_n &= a_{n1}F_1 + \ldots\ldots + a_{nm}F_m + b_nU_n \end{aligned}$$

where

Y_j = a standardized form of a variable with known data
a_{jm} = the factor loading or weight for each factor
F_m = a function of some unknown variables
U_j = a unique factor
and b_j = a unique factor weight

The variables entering into each function, F_m, are unknown and are related in unknown (and not necessarily linear) ways.

The factor analysis technique provides values for the constant, a_{jm}, called loadings which represent the extent to which each specific function is related to U_j. Once the factor loadings for each variable are determined, the initial set of statistics can be aggregated through the determination of factor scores into a single index, in which each variable is weighted proportionally to its involvement in a pattern; the greater the involvement the higher the weight. For example, an index, I_1, constructed from the first factor loadings may be expressed as:

$$I_1 = \sum_{j=1}^{n} \frac{a_{1j}}{\lambda_1} Y_j \qquad (5.7)$$

where λ_1 is the eigenvalue for the first factor. Blackman et al [7] suggest developing the innovation index from the factor scores for the first factor and that in no case does the examined first factor account for less than 53 percent of the total variance in the data.

Various simple methods for the determination of the innovation characteristics of industries or firms have been suggested or used. Innovativeness can be associated with particular industries or companies, firms, research organizations, etc. in a particular area of technology. While most innovation measures consider expendigtures or efforts in research and development of products and processes and the impact on sales, such indexes are usually applicable on either a macro basis such as a whole industry or to a very narrow specialized technology at the firm level.

Innovativeness furthermore is more a measure of ability to develop, apply, and introduce new technology than a measure of expenditure for innovation related R&D and other activities. R&D can be performed imaginatively and successfully or perfunctory. It may be aimed at innovation or just the discovery of new knowledge without transforming such new knowledge into applications and thereby technological innovations. Measures of innovation should therefore include not only factors which represent effort at innovation (in monetary or other terms) but also factors which indicate imagination, ability to identify or discover important technological voids, degree of sophistication of innovation process, recognition of physical and other constraints, judgement of market and other opportunities, development of an effective innovation strategy, and methods of approach to innovation.

For example, does the innovation process used take full advantage of technological developments in other fields? Is the innovation process properly planned to assure that decisions can be effectively made at each stage of research and development? Innovation should be considered a technology development project which must be properly planned. Such plans must include all possible outcomes from innovation activities. An innovation index could be based on the following quantitative and qualitative factors to be truly representative of the innovativeness exhibited, such as:

1. effectiveness of identification of technological voids and discovery of related technological developments;

2. planning of research and development of innovation;

3. projection of market and other opportunities and translation into innovation project objectives;

4. commitment of research and development resources (money, laboratory, engineering, etc. resources);

5. effectiveness of innovation strategy (elimination of duplication, use of related technology, etc.);

6. effectiveness of innovation organization and method of implementation; and,

7. timeliness of innovation process in terms of timing and speed of implementation.

While many of these factors cannot be determined explicitely, relative or ranked values for each and a weighing of the factors among themselves can be obtained by a 'Delphi' or other type of expert opinion survey.

5.4.2 *Use of the Innovation Index in a Determining Market Substitution Index.* We have found that the constant $\phi(t)$ which governs the substitution rate can be expressed as:

$$\phi(t) = Z + .530\ P(t) - .027\ I(t) \tag{5.8}$$

If the economic characteristics of historical technological innovations are known, such that estimates for P(t) and I(t) can be obtained and if the value of $\phi(t)$ is known from historical substitution rates, it is possible to estimate the value of Z. The magnitude of the Z values for a given industry reflect the propensity of that sector to innovate. It might be expected that a relationship would exist between the value of Z in an industry and the value of the innovation index for that industry (I_B).

This hypothesis has been tested by Blackman et al and it was found that such a correlation really exists and, indeed, the correlation coefficient was 0.92. The standard deviation of the regression analysis coefficient performed was .0615 and the following regression equation was obtained:

$$Z = 0.2221\ I_B - 0.316\ I(t) \tag{5.9}$$

So, the rate at which new markets can be expected to develop for new technological innovations can be estimated if the economic characteristics of new technological innovations can be estimated. By identifying the industrial sector in which the new innovation will occur, a value of the innovation index can be estimated from historical data and a corresponding Z can be found by using this relationship. The economic characteristics of the innovation can be used to estimate P(t) and I(t) and we can solve for $\phi(t)$.

Further research on the innovation index by Halliday and Lowitt [9] revealed that the innovation index has an underlying structure in which I_B values of the various industries are linked to integers, rather than being free to vary continuously. A factor analysis was performed with respect to four innovation variables:

- research and development expenditures, Year N
- planned research and development expenditures, Year N+4
- research and development as percentage of capital operating, Year N
- estimated new product sales percentage, Year N+4.

It was found that I_B can take on only integer multiples of an "innovation index measurement unit" and the values 0.26, 1, 2, 3 were observed.

The Value $I_B = .26$ can be explained by the fact that there is no viable industrial sector with zero potential for innovation for any real world industrial sector concerned with the survival of profits and markets, at least some of the subsectors or firms composing a sector, unique factors must come into play which can only serve to increase the measured value of I_B above absolute zero.

Essentially, there are two possible mechanisms which could generate integer-linking. Either a single independent variable, strongly dominant in the innovation process, can itself only take on integer values - and it is implausible to believe that any such variables would not already have been identified - or integer-linking arises from the collective behavior (i.e., dependent relationships) of all industrial sectors.

An analogy from biology will serve to illustrate the point. The species forming a particular ecosystem never displays the entire spectrum of conceivable responses to biological problems of reproduction, securing a food supply, etc., nor do they display a random subset of that spectrum. This is despite the fact that, if viewed only at a micro level, each species is free to evolve independently of the others. Instead, there exists a complex set of interspecies relationships and feedback looks at the ecosystem level. Relationships at this macro level define certain niches that can be occupied and species evolve to best fit one of those niches. Other questions also arise, such as which level of aggregation economic analysis should be made, etc.

The concept of innovation index, first formulated more than a decade ago, now appears ready to contribute significantly to a greater understanding of the innovation process. Needless to say, that is a topic of current and future research.

5.5 ORIGINS OF TECHNOLOGICAL CHANGE

The origins of discovery are often thought to rest with the identification of a need or a technological void, a chance observation, or the perception of a problem. Yet all these origins except the change observation are sometimes ill-conceived or structured, less because of the lack of knowledge of the important issues or facts then because of bias and preconceived ideas. Problem identification is not done scientifically, although the experiments performed to find a solution to a problem are usually scientifically designed.

Simon [8] in a recent study discusses scientific discovery and shows how the theory of problem solving can be used to explain some of the most important phenomena of scientific discovery. At the same time, there is evidence of the relevance of unconscious thought to the discovery process. Also many discoverers point to the importance of involuntary thought and reflection to the focusing and ultimate success of a discovery or problem solving process. Researchers have tried to develop models of the logic of problem solving and of discovery. Although the process of discovery can be studied, it is extremely difficult to develop a unique model of its logic or even the process of discovery itself, because most discoveries - even those instigated by the discovery or statement of a need or a problem - undergo different and often unique processes which could be rationally termed to be highly illogical. This may in part be due to the fact that problem statements are often ill-structured, but there are other reasons as well. One may be that a logical approach to the solution of a problem cannot be developed or is outside current scientific knowledge or capability. Another is often the lack of precision in the definition of the problem, and specification of what constitutes a solution to the problem. In many cases problem solution is by trial and error and not a formal logical step by step approach.

Logical solutions to problems including problems of discovery must rely on efficient solution seeking search, based on a continually updated data base which is designed to permit derivations not only of the expected outcome of any of the alternative solution approaches, but also the conditional probability that given the outcome is achieved it constitutes a solution to the problem. Obviously if the problem is ill-defined or badly structured, and/or the solution requirements badly defined, even an effective search and successful performance of selected alternatives leading to a solution discovery and invention may not provide meaningful benefits.

References

1. Arrow, Kenneth, "The Economic Implications of Learning by Doing", Review of Economic Studies, 29, 1962.

2. Hollander, Samuel, "The Sources of Increased Efficiency", MIT Press, Cambridge, Mass., 1965.

3. Myers, Sumner and Marquid, Don, "Successful Industrial Innovations", National Science Foundation, Washington, DC, 1969.

4. Klein, B., "Dynamic Economics", Harvard University Press, Cambridge, Mass., 1977.

5. Mansfield, E., "The Economics of Technological Change", W. W. Norton & Co. Inc., New York, 1968.

6. Dobrov, G. "A Strategy for Organized Technology", IIASA Publications, Luxemburg, Austria, 1978.

7. Blackman, A. W., Jr., Seligman, E. S., and Segliero, G. C., "An Innovation Index Based on Factor Analysis", Technological Forecasting and Social Change, Vol. 4, 1973.

8. Simon, H. A., "Models of Discovery", D. Reidel Publishing Co., Doidrecht, Holland, 1977.

9. Halliday, D. R. and Lowitt, H. Z., "An Integer-Based Blackman Innovation Index: Hypothesis, Evidence and Implications", Technological Forecasting and Social Change, Vol. 26, 1984.

6. Sources of New Technology

Technology can be acquired in different ways. It can be internally developed by R&D or discovery, it can be obtained by buying or otherwise acquiring an innovation resulting from R&D or discovery, during various stages of the innovation process, or after the technology has been largely perfected and approaches maturity. Methods of technology acquisition are quite diverse and include purchase, transfer, and licensing, which in turn may involve acquisition of knowledge, process, product, or right to use of the technology. Technology acquisition is complex because it combines efforts of protection of the rights of the owner and those of the acquirer.

Technology can not only be acquired at different stages of its own developments but also at different levels of a firm or organization and therefore, for purposes of strategy, productivity, or technology/market position. Technology acquisition can be implemented to achieve an incremental or large technological change. It can be introduced as a trial or hesitant move or as a radical change replacing other technologies under development or in use. Its purpose can similarly involve competitive or cost improvements, quality enhancement, new products or services, or a combination of such or similar goals.

Technology acquisition can be the result of or response to formal, tactical, or strategic analysis, competitive pressures, identification of short-term opportunity or failure, or belief in a new approach with resulting risk taking. The decision on if, when, what, how fast, and at what rate, and by what method to acquire new technology is handled differently by various firms and organizations, and depends on organizational, structural, cultural, historical, conditional, as well as personalities involved.

6.1 TECHNOLOGY ACQUISITIONS

Technology is acquired in order to reduce cost, improve quality, introduce new product (or process to perform an activity), or develop completely new activities or enhance or introduce new knowledge. Technology acquisition is often driven by a combination of such objectives and has as its ultimate goal improvement in the competitive, financial, and economic (which includes social) performance of the individual, firm, or organization. Technology acquisition can take place during the R&D, discovery or invention stage, the innovation stage, or the technology implementation or maturity stage. Methods of acquisition of technology depend largely on the stage at which technology is acquired and the personality/objectives of the owner and/or acquirer. During the R&D stage, the technological potentials are usually not readily definable and therefore the value of the technology is hard to determine. During the later stages of technology developments when benefits are more readily estimated, various formal economic approaches to technology acquisition are usually employed such as:

1. technology sale or purchase;
2. technology licensing;
3. technology limited rights licensing;
4. joint venture use of technology; and,
5. subcontracting.

The issue of economic incentives for and protection of technological developments is an interesting subject which raises questions about the role of licensing or other acquisition procedures in the process of technology development and innovation. In other words, does licensing or other economic opportunities enhance or deter the rate of innovation?

6.1.1 *Motivation, Rewards, and Incentives*. The rewarding of inventors or innovators is rarely 'fair' and is difficult in any case because of the problem of defining both the limits of the physical/intellectual property and the resulting property rights. Does an inventor, for example, have the rights of secrecy or complete protection which allows him to defer the use of an invention or innovation, when such use would greatly benefit society or solve an urgent societal problem? Or does an inventor have the rights of property of a discovery he made by change even on his own time, while fully employed, for example, as a government official, university researcher, or company employee. There are numerous examples of inventions where property rights are difficult to define or judge.

There are a number of ways management can motivate employees to invent and innovate, while maintaining the property rights of any ensuing invention with the firm. Common direct motivational approaches includes:

1. benefit sharing;
2. outright rewards;
3. prestige transfer by naming invention after inventor;
4. salary increase;
5. promotion (and project responsibility); and,
6. assignment of subsidiary right, such as rights to certain markets.

In many cases, management may offer deferred and non-monetary rewards or motivators to potential inventors or innovators.

There are a number of subsidiary questions though which may affect the approach. Does patent protection increase the incentives to invent and innovate? And does licensing inhibit the investment for innovation, both by the patent or license holder or outside investors?

There are diverse opinions on the question of the effect of the absence of licensing on inventors and innovators, and if the abolition would increase or decrease incentives to invent and innovate.

An increasingly interesting question is the motivation and incentives of transnational licensing, which is often more concerned with the efficiency or effectiveness of technology transfer than with market control as described in later sections. There are also many non-monetary incentives for inventors. Advance or greater insight in knowledge, drives some inventors and even some innovators.

Another incentive is curiosity which often leads to chance discoveries which constitute inventions. Pride and prestige are other non-monetary forces driving inventors. The importance and potential of monetary rewards increases during the innovation stage and the perception of estimates of their magnitude often affects the level of financing and thereby the rate of progress of the innovation process.

6.2 TECHNOLOGY TRANSFER

Technology transfer is an increasingly common method for the acquisition of technology used by firms or even nations. Technology transfer can occur as a commercial transaction, as a result of bilateral agreements between firms or nations, as a unilateral transaction, or even sometimes as an involuntary transfer of knowledge and technology. There are numerous avenues for the transfer of technology and many methods for the actual transfer process. For technology transfer to be effective, the transfer must be effectively planned and assurance established that there is:

1. need for the technology;
2. absorption capacity for the technology;
3. availability of resources for the technology transfer;
4. proper social, economic, and technical environment for the technology; and,
5. appropriate factor prices for effective use of technology.

Technology transfer is the method whereby technological knowledge or physical technology is transferred from one firm to another within or without the same country. Although technology transfer is usually associated with the introduction of technology to developing from more advanced countries, we will consider here the broader definition of technology transfer.

The technology transfer process consists of need identification, technology assessment, technology transfer planning, and technology transfer implementation. These four steps are shown in Figure 6.1, and are usually required to assure effective selection and introduction of a technological change.

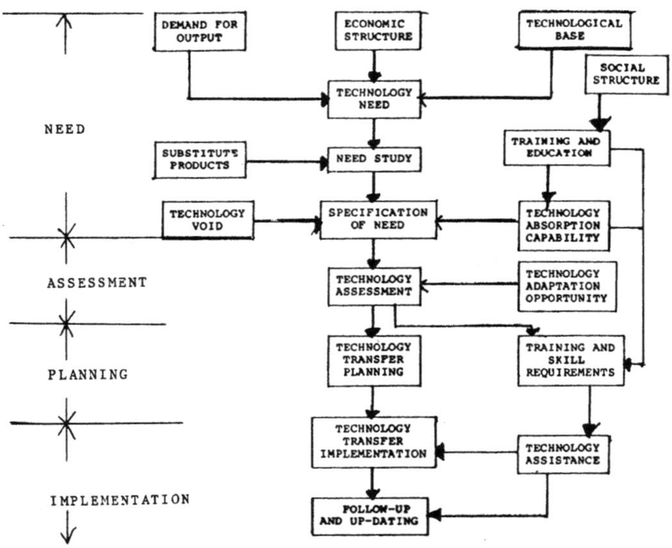

Figure 6.1. Technology Transfer Process.

The <u>need identification</u> usually involves an evaluation of the existing technology and economic needs, market demands, competitive factors, use of both existing technology, its output and the output of potential replacement technology, the technology base and environment that exists, the complementary technologies available (locally and externally), and finally the resources available (human, knowledge, material, financial, etc.).

<u>Technology assessment</u> is concerned with the identification of all existing or under development technologies which do or could fill the identified need. Assessment includes

evaluation of status, performance, accessibility, and acceptability of the alternative technologies identified. In technology assessment we also consider the trend of technology development, including that of competitors using in-house or other sources. It also includes study of sector involvement, adaptability, diffusion, and substitution effects, particularly if the new technology is expected to replace existing technology.

The next step - technology transfer planning - is concerned with:

1. acquisition process planning;
2. financial planning;
3. physical and/or knowledge transfer planning;
4. market planning;
5. implementation planning;
6. operational planning;
7. skill and acquisition training;
8. maintenance; and,
9. management.

Technology transfer planning is an integral part of technology assessment and contributes to decisions affecting the choice, timing, and method of acquisition of technology.

Finally, the technology transfer implementation stage consists of a number of distinct steps which are usually undertaken:

1. evaluation of alternative ways of introducing a new technology into the historic, indigenous technological gap;
2. selection of method of chosen technology introduction;
3. identification of technology framework and requirements for introduction into setup, including absorption, adaptation, linkage, and staffing;
4. prototype testing or simulation as well as operational planning;
5. commercial development and resource flow planning;
6. scale development;
7. development of training and management system;
8. setup of operating procedures;
9. development of technology maintenance and support system; and,
10. establishment of controls.

Technology transfer can be direct or brokered; timely or delayed, and continuous or discontinuous. The timing of transfer obviously affects the built-in obsolescence (or lack thereof) of the technology selected. Technology transfer decisions can similarly be made in isolation or with full participation of those involved in benefitting. In other words, such decisions can be imposed on the ultimate user of the technology or involve him in the whole technology transfer process.

When technology transfer does not live up to expectations, the reasons are generally found to be the result of:

1. lack of understanding or effective evaluation of the need and problem the technological change is to address, including the environment in which it is to work;
2. preconception of solution to the assumed need or problem;
3. overestimation of absorption capacity of the environment;
4. setting of unrealistic, irrelevant, or inapplicable objectives of standards;

5. lack of development of measures of evaluation;
6. lack of complete cooperation between donor and recipient;
7. lack of follow up;
8. lack of continuity of technology transfer process;
9. insufficient resources;
10. excessive delays in technology transfer process;
11. lack of adequate maintenance;
12. inadequate training;
13. incompetent management; and,
14. inappropriate use of new technology.

Adaptive or appropriate technology are often concepts used in technology transfer, particularly when applied to the introduction of advanced technology into developing countries.

The question seldom answered though is what is implied by adaptive or appropriate technology. There are many cases where technology transfer was based on over- or under-assessment, technology dumping, and transfer of captive technology (with strings). In other cases, it was tainted by political considerations. At other times, technology is introduced because decision makers are more interested in the prestige than appropriateness, usefulness, and effectiveness of new technology.

As import of technology by developing countries consumes a significant percentage of their GNP (10-19%), technology transfer must be planned as an integral part of development policy. It must include effective resource planning and allocation and the risks and uncertainties must be understood (or explained). Technology transfer must be considered a long term and continuous process and not a periodic or stepwise process. Institutional and cultural constraints must be evaluated before considering a technology transfer, and the role of vested interests must be identified before commencing the technology transfer process.

Economic evaluation of the effects of technology transfer should include not only the traditional costs and benefits associated with a process or product but also

1. transfer costs
2. exchange risks
3. training and knowledge acquisition costs
4. technology maintenance costs and risks, and
5. technological performance risks.

Similarly the objective used to weigh decisions affecting technology transfer include

1. strategic objectives
2. financial/economic objectives
3. political objectives
4. moral objectives, and
5. social objectives.

The impact of technology today is so broad and pervasive that the technology transfer process must be subjected to more than purely technical and economic goals and resulting decisions.

6.2.1 *Effectiveness of Technology Transfer Directed Toward Less Developed Countries.* Technology transfer and its effectiveness has been a subject of lengthy discussions for many years, yet there seems to be little real understanding of the methods and approaches which should be applied to assure success in such transfers, particularly when the transfer is between industrialized and non-industrialized nations. While technology transfer among industrialized countries is usually a matter of commercial expediency, sale of knowledge, transfer induced by pressure of competition, or transfer by chance; technology transfer between industrialized and developing countries is often implemented under unilateral or bilateral technical, strategic, defence, or economic assistance agreements, and nearly always involves government - at least on the side of the developing country. Technology transfer has become a major tool for gaining or maintaining commercial influence, political or strategic leverage, or achieve some economic advantage, by both or either technology provider or technology recipient. Today it is an area of intense international, and particularly, East/West competition. There are serious indications though that among Western countries the U.S. is not doing as well as its technological stature and development aid contributions and expenditures warrant. Many claim that the methods used and technologies selected for transfer to developing countries are often inadequate or inappropriate.

The purpose of this chapter is to evaluate the reasons for the apparent lack of success in technology transfer to developing countries, both rich (OPEC) and poor, and discuss the role of marketing and competition in technology transfer, ownership and government role in the organization and control of the technology transfer process, and finally the selection of the methods applied in the technology transfer, and training of users of the technology transferred.

An important element in successful transfer of technology is the objective and accurate assessment of the effectiveness of technology transfer. It is found that in the majority of cases little, if any, post transfer assessment is performed, and therefore little is learned from the experience of earlier attempts of technology transfer.

Technology transfer assessment is generally interpreted as a method of policy analysis which systematically defines, explores, and evaluates direct and indirect economic, social, environmental, and institutional consequences of the introduction of a new technology into a community or of expansion of existing technology within a community. The purpose of technology transfer assessment is to evaluate the effectiveness with which technology transfer has been, is, or will be achieved. Measures of effectiveness of technology transfer vary widely and can usually not be defined explicitly.

The quality of a technology transfer assessment is not the sole determinant of its usefulness. Usefulness, particularly as a measure, is also a function of contextual factors which are largely beyond the control of the analyst. The major contextual factors are: political, economic, social, and organizational.

The political context is readily apparent in most technology transfers. The large majority of technology donors or producers and potential users believe that technology assessments have political implications. Perhaps the most striking example is the relationship usually found between perceived treatment by the evaluator and the existing or potential user. User organizations generally believe strongly that technology transfer assessments are not in their interests, and are designed to eliminate shortcomings on their behavior, organization, or use of the technology transferred. Individual factors also bear on utilization. The longer the tenure of a user in an organization, or the longer the process of technology transfer has gone on, the less the utilization technology assessment. The prior knowledge of the capabilities, characteristics, and other factors of potential users also bears on utilization of technology assessment. Utilization of technology assessment is often higher among those with little expertise in the concepts of technology assessment than among the better-informed. Yet, those most familiar with the field of technology assessment do make greater use of particular technology transfer assessments. These general conceptions appear inconsistent. This though is not the case as users able to understand the rationale of such assessments do not necessarily have any expertise in such assessments. With respect to organizations, it was found that technology assessments were most likely to be used by those which consider their function important and long term, engage in long-range planning, and who would consider making decisions relevant to the issues addressed in a technology transfer assessment. One problem is usually the degree of acceptance of information and findings produced outside the organization. An important factor is the aversiveness or lack of aversiveness to risk by an organization. Risk here implies failure (real or perceived). It is therefore important to include risk and uncertainty in the assessment of technology transfer. Such risk and uncertainty is involved in many correlated and uncorrelated factors.

As noted, the technology transfer process is complex, often ill defined, and usually subject to many problems. Technology transfer has generally been considered a stepwise or discrete process which assists Less Developed Countries (LDCs) to develop their economies and which permits a gradual convergence in per capita income and standards of living between developed and less developed countries. The results of technology transfer are quite diverse. In general though technology transfer has not succeeded in providing a meaningful convergence in economic terms. There is risk and uncertainty in maintaining technology transfer flow under continually varying political and economic conditions. Similarly institutional constraints and incentives are often fluid. There is also the question of the advantage of stepwise versus gradual continuous technology transfer. Most technology transfer comes in spurts, with large programs or projects in a sector followed by long periods of technology transfer inactivity. Little, if any, feedback information on technology use in the new environment is usually available, and feedforward information is generally found to be inadequate because of the long time interval between projects with meaningful technology transfer requirements.

To succeed it would be better to make the technology transfer process continuous and dynamic with constant feedback and feedforward of information and knowledge or experience flow. This would permit more effective technology transfer planning, better distribution or use of limited aid or investment resources, and better selection of appropriate technology and institutional as well as manpower resources. Appropriate technology as used now is an expression which begs the important question of gaps in LDC development. Appropriate technology must be related by reference to development criteria and experience in technology transfer and use. This can only be accomplished if technology transfer becomes a dynamic continuous process with feedback and feedforward of information as described. Development criteria must be specific enough to permit effective selection of real time requirements of technology transfer, yet they must not be overspecified as this would constrain technology and resource selection.

It is important to note that, to be appropriate, technology need not be an optimum technology in terms of a particular need. It should be the best technology to meet a specific or general development plan and the related criteria for success. Feedback and feedforward is also required because changes in development affect level and structure of incomes, which in turn affect factor prices as well as the patterns of product demand. This in turn may affect decisions about the choice of appropriate technology and distribution of available resources. Dynamic changes in world trading patterns similarly affect import and export possibilities which should be considered in the choice of an appropriate technology.

Notwithstanding the massive efforts and expenditures for in technology transfer, the process has generally been less than effective. There are a variety of reasons that can be advanced to explain some of the shortcomings. Uppermost is probably the large gap in wealth and technology between developed and developing countries. This fact which is the main reason for the need for technology transfer is also a major cause for distrust. The relation between donor or supplier and recipient of technology is one conceived of unequal partners. There is also a perceived concept that there are factors within the western industrialized economies which tend to systematically distort all choices involving technology transfer.

Policies regarding technology transfer by developed countries are usually not aimed at the particular development needs of LDCs. Many LDCs today are governed as planned economies. Yet many lack the basic skills or resources for effective top level planning. Little use has been made of intra-LDC technology transfer, whereby experience gained in similar economies is transferred and a continuous flow of technological advance assured. Such flow could be organized in multiple directions depending on the level of technology in particular sectors of various LDCs.

Unfortunately many LDCs themselves want to make quantum advances in technology, however unreasonable or unachievable, instead of opting for steady, continuous improvements which are more readily absorbed, accomplished, and financed. This spurious ambition for the introduction of the most advanced technology is probably the major cause for the lack of success in technology transfer. It is unfortunate that most developed nations have acceded to this fallacy for political reasons or reasons of national or personal ambition, at tremendous cost to them and, more importantly, the recipient LDCs.

A major problem is often the lack of followup or post transfer analysis. There is a need to continually evaluate the effectiveness of technology transfer both with respect to the purpose of the particular sector in which the technology was applied as well as with respect to the overall development plan and environment of the LDC. Furthermore the evaluation must include consideration of changes in the external environment as well as to assure that the continuous trend of technology transfer to an LDC is in line with changes in both the internal and external environment. Only in this manner can we assure that technology transfer will become more effective, and make a meaningful contribution towards closing the gap between rich and poor nations.

For technology transfer to a developing country to be effective, it should form part of national, regional, or local development policy, otherwise it forms a small, disjointed technical anomaly which soon disintegrates. It should also include effective resource planning and allocation so that follow-up requirements of the new technology, such as spares, are properly provided for. Risk and uncertainties involved in the technology transfer must be understood and prepared for. Such risks include the risk of environmental (society, labor, etc.) acceptance of the change.

To be effective, technological transfer to LDCs must be long term and not just a one-time stepwise process. Finally, it is important to evaluate institutional constraints as part of technology transfer planning to assure their resolution or acceptance in the transfer plan.

6.2.2 *Technology Transfer Pricing*. In technology transfer, the 'perfect competition' market mechanism of pricing does seem to be relevant or to apply as the 'commodity' for sale may well be unique. The numbers of sellers and buyers, similarly, may be limited to one on each side and information on the 'technology or commodity' of the transaction may not be widely distributed. A price that is based upon the cost of Research and Development and innovation (if undertaken) often serves as a lower bound for the price, though the potential use by and value to the seller is usually a more important measure of price, as the price of a technology should reflect the present value of the future cash flow resulting from its implementation rather than its cost. As noted before, technological development goes through three stages of a logistic curve (Figure 6.2) which represents the technical life cycle of R&D (discovery), innovation, and maturity when general diffusion takes place and the technology is approaching obsolescence and ultimately substitution.

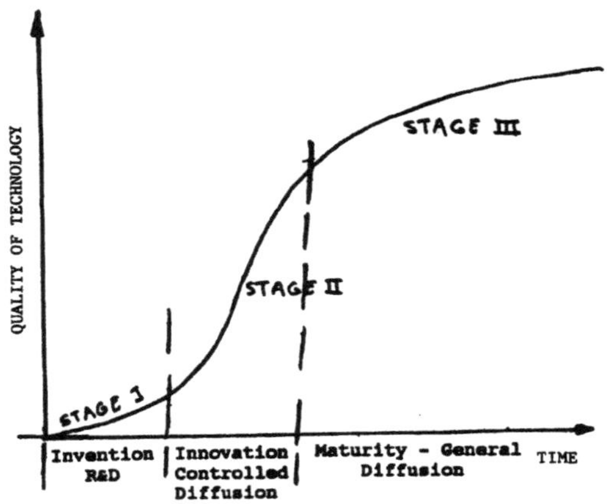

Figure 6.2. Evolution of a Technology.

R&D is the least structured stage because of the great deal of risk involved.

The second stage is usually characterized by a rapid development of the technology. This rapid growth may lead the owner of the technology to very high expectations with regard to its future. This, in turn, may result in an unrealistic high price, and a very concrete risk is also attached to the deal. The transfer period, which can be defined as the time that passes between the definition of the technological package to be traded and its application in full-scale production, may last several years. During this transfer period, the seller has the opportunity to develop its own technology quite considerably and create a new advantage over the buyer.

In the third stage, the transfer of technology takes place with lower conflict of interest between the seller and the buyer(s). At this stage, considerations of technological maturity and rate of obsolescence when overtaken by more effective, newer technology must be taken into account. The seller may regard the technology transfer as a marketing tool. The seller will also consider alternate uses or sales of the technology in the future. The planning horizon of the technology transfer decision is determined by the estimated life span remaining for the technology. In general, pricing decisions in technology transfer are affected by:

1. size and market share of firm which owns the technology, including its ability to actually use or diffuse the technology. Small, foreign, and science (not production or marketing) based firms are a priori cheaper sources of new technology.

2. size of economic unit of technology developer. For example an inventor in the U.S., Europe, or Japan has greater ability to market a new technology.

3. maturity of firm developing technology.

Most developing countries procure or obtain new technology which is either insufficiently developed and marketed in developed countries or more frequently which is quite mature and approaching obsolescence. As developing countries require mote time to introduce a "new" technology, they do, as a result, slide easily into technological obsolescence even with newly acquired technology.

6.3 TECHNOLOGY LICENSING

Technology licensing is the sale of information and rights to the use of new technology. It usually occurs during the technology development or innovation process, although it may also happen during the initial discovery or during diffusion during or at the end of the innovation stage when the technology is well developed. Licensing consists of formal contracts which specify the rights and duties of the licensee and licensor. The licensee usually acquires access to knowledge about the technology, including training if applicable, as well as rights to the use of the new process, product or service technology. A license often stipulates the limitations on the use of the knowledge or technology acquired.

Market, geographical, sectorial, quantity, and quality constraints are usually imposed by the licensor, as well as limitation on the diffusion, publication, sale, or other disposition of the knowledge or use transfer to non-authorized parties. Licensors generally attempt to control all diffusion and adoption of the new technology, at least for a stipulated period of time, and/or region. They may similarly restrict use to particular application or groups.

The licensor, on the other hand, undertakes to provide the most up-to-date knowledge and application of the new technology and often to train or instruct licensee personal in its use.

There are several incentives for licensing for both the licensee and the licensor. The licensee expects to obtain more advanced technology than in use by him (and/or his competition) at lower cost and in shorter time than it would take him to develop similar technology, if at all feasible. He also reduces his risk, by purchasing developed knowledge and application which is at least partially, if not completely, proven to work and to be an advance on existing technology in use. Similarly it allows more effective timing of technological change in process or product based on market and productivity or cost trends than the acquisition of new technology by in-house research and development.

The licensors' objectives in selling licenses are largely to get help in financing the development of the technology and share the risk of technology development and application with others. When the new technology is advanced and has large market and profit potentials, licensing may not only permit recovery of research, development, innovation and other improvement costs, but may also allow large profits to be obtained. Furthermore licensing may discourage potential competition from investing in competing research and development. This though may be affected by the cost of the licenses in relation to the real or potential advantages of the new technology and projected research and development cost. For example, involvement of a powerful licensee may influence financial commitment by others.

Licensing affects the development of new technology and may encourage or discourage new research and development. Most product or process technology licenses require payment in the form of royalties for the use of the new technology. Royalties are determined on the basis of the cost and market advantages the new technology offers in relation to an existing technology. They are also influenced by the state of development, the maturity, and probable obsolescence of the new technology. In other words, licensees will consider not only the cost and market advantages a new technology may provide, but also its state of development (or reliability), its state of diffusion, its potential for further improvement (learning), and the risk of a newer technology invasion within a time too short to fully use the new technology in terms of cumulative output and profit or cost savings.

The licensor will consider the advantage the new technology provides in the pricing of the royalties or other payments and the timing of licensing. The incentives for licensing by the licensor are greater when the advantages of the new technology over the existing technology it replaces are small and smaller when the advantages are great, unless the licensor cannot continue the technology development with licensees.

Licensing is therefore a strategic decision for both licensee and licensor, and requires effective market, technology, and cost valuation and forecasting so as to develop an effective strategy of timing and pricing of technology licensing by the licensor, and of choice, timing, method of application, and benefit objectives by the licensee.

The terms of licenses can be quite varied. The may give the licensee wide ranging rights to the use, sale, and access to future developments of the new technology, or may be quite restrictive. Some restrictive licenses affect not only quantity, quality, form, and application of the new technology, as well as geographic area or market segment in which the new technology can be sold, applied, or used, but even may require the licensee to keep knowledge of and the use of the new technology secret.

A problem researchers and developers often face is that the product or service the new technology represents, has a short economic life, independent of its stage of development, and that the price that can be charged for a license depends on the remaining value of the new technology in terms of the advantage over existing technologies for the remainder of its economic life, which comes to an end when the technology is overtaken by more advantageous technologies. The advantage of an invention is therefore a dynamic measure which can erode very quickly and thereby erode the valued licensing of the invention or innovation.

The price obtainable for a license to use a technological innovation usually peaks before the end of the innovation period as shown in Figure 6.3. The problem innovators face is therefore the determination of this peak, whose position is affected by

1. relative advantage of new technology in terms of cost (profit) and quality in relation to existing technology,
2. market size and potential,
3. competitive position or market share of licensee,
4. state of development of other newer technologies with similar or substitute applications, and therefore the projection of the remaining economic (competitive) life of the new technology,
5. state of development of new technology and complexity of new technology, and
6. rate of innovation of the technology (fast or slow rate of innovation).

In many cases the rate of innovation is closely affected by the sale of licenses when the licensor attempts to finance much of the innovation by licensee fees or royalty payments. It is important that the licensor does not only assure effective timing of the sale of licenses, but also determines the most effective size and expected use (and development) of the new technology by the licensees. It may then be advantageous to sell the initial or early licenses not to a large, powerful firm capable of using the new technology to rapidly improve their costs by learning, but to make the initial license sales to small users so as to keep the advantage of the new versus the existing technology larger, at least until a later stage in the innovation process when peak prices for licenses can be obtained. Developing an effective licensing strategy can be considered a dynamic gaming exercise which, at some stages, results in a zero sum game. A basic strategy payoff matrix for consideration by a licensor is shown in Figure 6.4. The licensees will develop their strategies on the basis of the negative of the payoffs of the licensor, though the payoffs among the licensees may vary with size, market share, cost of production or product, slope of learning curve, and time since introduction of last technological change.

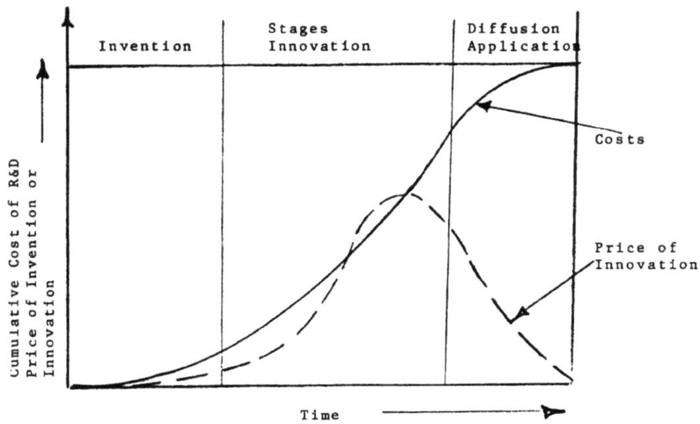

Figure 6.3. The Development of Price Obtainable by Licensing versus Cost of R&D/Innovation.

		Licensee 1	2	3	4	5		
	1	P11	P12	P13	P14	P15	Early	Exclusive
	2	P22	P22	P23	P24	P25	Early	Non-exclusive
Licensor	3	P31	P32	P33	P34	P35	Middle	Exclusive
	4	P41	P42	P43	P44	P45	Middle	Non-exclusive
	5	P51	P52	P53	P54	P55	Late	Exclusive
	6	P61	P62	P63	P64	P65	Late	Non-exclusive
		Early Exclusive	Middle Exclusive	Middle Non-exclusive	Late Exclusive	Late Non-exclusive	Timing of Sale of Licenses	Term of Licensing

Figure 6.4. Licensor's Payoff Matrix.

In the basic one-stage, two-dimensional strategy model shown, the licensor and licensee determine the value of their respective alternative strategies for the selling or buying of the invention or innovation. For simplicity alternative strategies for both players are assumed to differ only in the timing of the licensing (early, middle, or late in the innovation process) and in the fundamental licensing terms (exclusive or non-exclusive). Obviously in a real world example many additional strategies would have to be considered, including timing, term, and pricing alternatives which can be expressed as distinct strategies. In addition, the expected payoffs p_{ij} which assume the licensor assumes a strategy i when the licensee assumes a strategy j will usually be time varying when it is expressed as $p_{ij}(t)$. The payoff may not only be time varying, but also probabilistic, particularly if the payoff is largely based on future use and value (royalty or similar future payment conditions).

The market for licensing is quite imperfect in economic terms. Licensee/licensor negotiations can be termed a non-cooperative game, in which each player maintains his strategies as well as his expected payoffs secret. A licensing strategy gaming model therefore is only an attempt to optimize the players (licensor or licensee) outcome in terms of assumed payoffs, and not in terms of the payoffs of the opponent who, while theoretically confronted with the negative of the players payoff, in reality often faces unexpected and newer strategies and resulting payoffs by other players. An example of licensor/licensee strategy development is shown in Appendix 6A. As noted, licensor and licensees can obviously assume mixed or pure strategies, and strategies may change over time.

Although it is usually assumed as discussed in the next section that patent protection is a necessary condition for licensing, there are numerous examples of licensing without protection by patenting. A license may for example be sold by an inventor before completion of the research and therefore before a patentable discovery as a conditional license to use the results of the discovery when developed. This often includes participation by the licensee in the research or innovation. Most license contracts though are entered after patent coverage is obtained [1].

Licensing against regular payments based on the volume (quality) or revenue obtained from the use of the invention or patent is the most common form of commercialization of a discovery, though the licensor runs the risk that the license may not make effective use of the license. In fact he may enter into an exclusive licensing agreement with the intention of removing the invention or new technology from the market so as to maintain a monopoly, particularly if the existing technology available to him, though inferior in long term productivity or costs, requires no investment, no retraining or relocation and firing of workers currently employed. A licensee may also have other reasons, such as large investments in inventory of marketing which make it attractive to him to retain use of the old technology.

Outright sale of an invention, innovation, or patent involves fewer risks because it is independent of the licensees use, but may be less rewarding because the inventor/innovator does not participate in the user's success [2].

Most licenses today are non-exclusive and often restrict use of the new technology in terms of quantity produced or sold, geographical area of production or sale, and market sector. Licensing primarily involves public corporations, many of which maintain extensive inventories of license agreements and often exchange or sell licenses if the license permits. Inventors often resort to legal action when a licensee makes little or no use of an invention and simply removes it from the market.

6.4 PATENTS AND OTHER METHODS OF PROTECTION OF INVENTIONS

A patent is a right conferred upon an inventor to the exclusive use of his invention. This right which an inventor can trade under most patent laws in a way gives an inventor powers of monopoly over his invention. Patents are the most important method of protection of inventions, although not the only ones. They are usually justified as the most important incentive for inventors, as they provide him with a lawful opportunity to benefit from the invention as well as control over the methods whereby knowledge of the invention is diffused and the invention is utilized. A patent allows the inventor to withhold knowledge of the invention, delay or prevent its use, and control any and all rights to it. There are many who feel that patent laws are against the public interest by limiting and/or delaying use of inventions, many of which may serve to improve public well being. At the same time, it is recognized that inventors require incentives to invent and firms incentives to invest in cost research which may lead to inventions. Patents and similar restrictive or protective measures appear at this time to be the most effective incentive available.

To deter patent holders from preventing or withholding use of an invention whose use is in the public interest, proposals have been advanced to restrict the rights of patent holders and require them to use or issue licenses for use of an invention within a reasonable time. Patents have been issued in the U.S. since 1790 (or nearly 200 years ago) when Congress passed the first American Patent Law. The first official letters patent law though was introduced by the British Parliament in 1623 or nearly 167 years earlier. The British law awarded to the 'first and true inventor of a process' exclusive rights for a period of 14 years. The U.S. law, built upon the British example, confers the rights of exclusive use to the inventor for a period of 17 years. Nations interpret the rights conferred by patterns differently. While some consider these similar to the natural right to property, others do look at it as a special privilege given to the inventor by the government.

The average number of patent applications in the U.S. has grown from about 50,000/year in 1901 to over 100,000 in 1987, or a 2.2 fold increase. The percentage of patents issued (63,000 in 1986) to number of applications per year (110,000 in 1986) has remained reasonably constant at about 60% during the last 87 years. It is interesting to note though (Figure 6.5), that the percentage of patents issued to individual inventors has fallen from 81.7% in 1901 to just 19.2% in 1986. At the same time, the percentage of patents issued to U.S. corporations has increased from 7.1% in 1901 to over 42% in 1986. Patents issued to others, mainly foreign corporations and individuals, has increased from just over 1% to nearly 40% now. In other words, nearly 40% of all U.S. patents are now being issued to foreign individuals and firms.

Legal requirements for the issuance of patents vary among nations, as does the time to obtain a patent, length of patent protection period, and extent of legal protection under a patent. While it takes only 1-2 years in the U.S. to obtain a patent, from the time or original filing of a patent application, it takes as much as 5-7 years in other countries, where the search for patent conflicts is sometimes more stringent. For an idea to be patentable it must usually be

1. new and not obvious
2. useful
3. not previously achieved or used, and
4. safe as well as in the public interest.

These requirements differ somewhat among nations, and the interpretations of requirements are also broader or narrower. Similarly safety may mean not injurious to producer, user, or may include the total environment and the public at large. The patenting process normally includes:

1. <u>Specification</u> of the invention, its design, and method of use.

2. <u>Claims</u> made as to the particular advantages, benefits, and improvements introduced by the invention.

3. Search - In the U.S. the claimant or his representative assume the responsibility of proof that the patent filed and claims made are original and do not infringe on the rights (and claims) of any other patents filed or inventions with other right of priority.

4. A Notice of Allowance is given to an applicant whose claims are found valid and whose search and check by the Patent Office is found not to indicate any conflict with existing or prior rights.

To prove priority in case of conflicting claims interference is usually disproved by a test of date of conception or reduction to practice which then becomes the criteria of a patent award. There are many arguments for and against the maintenance of a patent system which can be summarized as follows. There are proponents and opponents to the patent system.

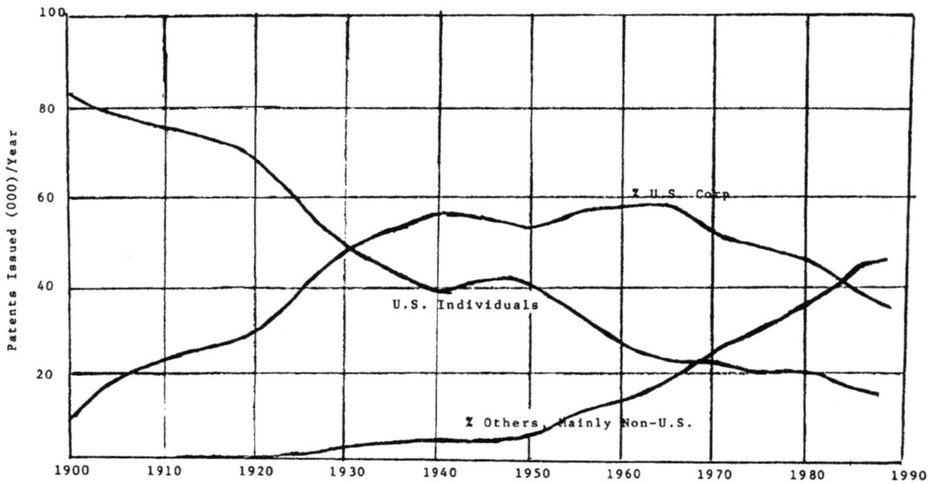

Figure 6.5. Percentage U.S. Patents Issued to U.S. and Other Inventors.

Arguments for Maintenance of Patent System

1. Property rights must be given to inventors to assure disclosure and inventive processes. Only this way will proper motivation be maintained for research, development, invention, and innovation.

2. Assured patent rights assure disclosure and thereby contribution to knowledge.

3. Patent protection is essential to generate investment for R&D and innovation.

4. Patent protection assures a continuous drive towards improved productivity and products.

<u>Arguments against Maintenance of Patent System</u>

1. Patents create monopolies.

2. Patents encourage suppression of invention by patent holders.

3. Patents cause others to waste resources to invent or develop solutions which circumvent existing patents.

4. Patents cause large legal costs and delays in the introduction of new technology.

5. Other motivations such as prestige are offered by modern society.

6. Marginal cost of all information should be zero.

It is generally assumed that licensing raises the incentives for the development of an invention. Similarly licensing helps spread use of superior technology and lower development costs and thereby unless oligopoly should help to also lower price, as different parties to licensing have competing objectives.

The opposition to patents and licensing is caused in part by the increasingly frequent non-use of patents by bold inventors or purchasers who often use patents not to protect their use of an invention, but to prevent others from using an invention which may successfully compete with their current products or services. This is particularly the case where markets are inelastic.

Patents basically provide the holder with a monopoly for the use of the invention for a given period of time (17 years in the U.S.). Patents rights can be transferred by

1. outright sale,
2. deferred sale,
3. partial (or geographic) rights sale, and
4. licensing against
   a. royalties
   b. periodic payments
   c. share of profit
   d. share or percentage of sales, and
   e. fixed percentage of output.

In economic terms, if a patent holder tries to maintain a monopoly, he would attempt to maximize his short run profits by producing a quantity Q* (Figure 6.6) at which his marginal costs equal his marginal revenue, but if a competitor enters the market he could force a lowering of the price to say P_c at which marginal costs are equal to P_c on the demand curve and a quantity Q_c at which average total costs are C. This would result in a significant reduction of the profit to the patent holder, particularly if his average total costs - which include developments costs - are significantly higher than those of his competitor.

In conclusion, the sale, use, or non-use of patents is a strategic decision which usually benefits by a formal analysis as a non-cooperative strategic game, as the market for licensing or sale of patent rights is imperfect (a non-zero sum and non-cooperative game).

The timing and choice of a patent rights sale or use decision depends on

1. the relative advantage of the new technology,
2. the state of development of the new technology,
3. capability and state of competing technology,
4. rate of change of new technology - rate of innovation progress,
5. the size and value of the market for the new technology, and
6. the competitive position of the new technology.

The basic concepts of game theory are presented in Appendix 6A.

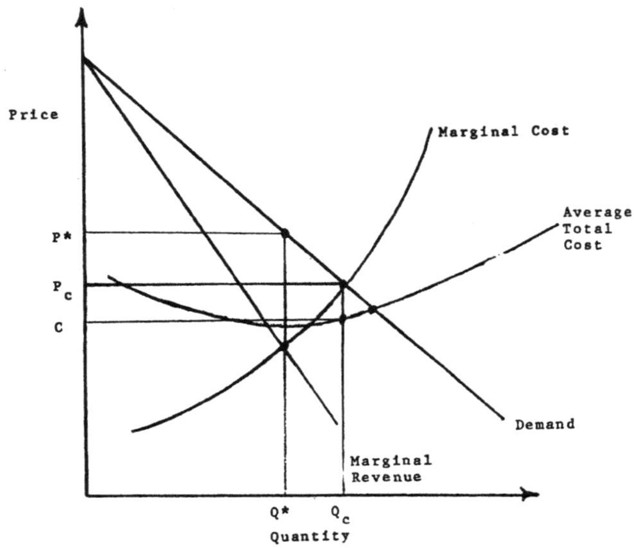

Figure 6.6. Impact of Price Reduction of Competition to Monopoly of Patent Holders.

6.4.1 *Patent Protection.* Patents, as mentioned, are designed to permit the inventor or the investor who acquires the rights to a patent to profit from the invention in any lawful way by providing them with protection against the use or sale of the invention by others without the explicit permission of the patent holder.

While patents in the U.S. provide the legal protection against unauthorized use, patent holders must use every means available to protect their rights under the patent. These means may require suing in court, challenging of patent infringer out of court, or various other proactive approaches designed to assure the commercial value of the patent. Unless potential infringers are promptly challenged, patent holders may loose valuable commercial benefits as well as rights under the patent.

Infringement action should consider the objectives of the patent holders. These may include

1. collection of monetary damages,

2. prevention of use of patent by infringer directly or indirectly,
3. transfer of infringers market share obtained unlawfully through patent infringement,
4. prevention of use of patent related or consequential ideas, and
5. punishment (or ruin) of infringing competitor by imposition of punitive damages.

Most patent disputes are entered with one or more of these objectives in mind. As the consequences of such a dispute can be devastating, patent dispute strategies must be chosen carefully. Strategies may be offensive or defensive to assure optimum competitive advantage under the prevailing legal rules offered by the patent system.

It may be advantageous to offer settlement of a patent suit in some cases, while carrying it through initial to appeals courts may be more effective in others. The most important issue in patent disputes is timing, and in case the suit is brought by the patent holder, as it usually is, it is essential to introduce preliminary or early injunctions to prohibit the assumed infringer from continuing to produce, use, or market the protected invention, and thereby enlarge his market share which may cause irreparable damage to the patent holder. These can seldom be corrected or paid for out of court imposed damages, as the infringer may not be insured and his own assets are often inadequate to cover the damage. Patent holders must be on constant watch for patent infringement, however minor, as inaction may be interpreted as lack of diligence or basic agreement with the infringer's actions.

References

1. Gallini, N. T. and Wrater, R. A., "Licensing in the Theory of Innovation", Rand Journal of Economics, Vol. 16, No. 2, 1985.

2. Katz, M. L. and Shapiro, C., "On the Licensing of Innovations", Rand Journal of Economics, Vol. 16, No. 4, 1985.

Bibliography

Adelman, I. and Thorbecke, Ed., "The Theory and Design of Economic Development", The Johns Hopkins Press, 1969.

Baranson, "Industrial Technologies for Developing Economics", Praeger, 1969.

Behrman, Jack N. and Wallender, Harvey, "Technology Transfers to Wholly-Owned Affiliates: An Illustration of the Obstacles to Controls". Paper presented to International Studies Association, Toronto, Ontario, 1976.

de Cubas, Jose, Technology Transfer and the Developing Nations, Council of the Americas and Fund for Multinational Management Education, New York, 1971.

Didijer, S., "National R and D Policy as a Social Innovation", in Management of Research and Development, OECD, Paris, 1972.

Ellul, J., The Technological Society, Knopf, 1964.

Evenson, Robert, Technology Generation in Agriculture, in Lloyds Reynolds (ed.), Agriculture in Development Theory, Yale University Press, New Haven and London, 192-223, 1975.

Ferne, G., "In Search of a Policy", in Research System, Vol. I, (France, Germany, U.K.), OECD,

1972.

Harper, P. and Eriksson, B., "Alternative Technology: A Guide to Sources and Contacts", Undercurrents No. 3, 34, Cholmley Gardens, Aldred Road, London, NW 6, 1972.

Havemann, H. A., "The Adaptation of Technology to the Situation in Developing Countries".

Hawthorne, Edward P., Report of OECD Seminar in Istanbul 5-9 October 1970, The Transfer Technology, OECD, Paris, 1971.

Helleiner, G. K., "Manufactured Exports from Less Developed Countries and Multinational Firms", Economic Journal, Vol. 83, 329, March 21-27, 1973.

ILO, Scope, Approach, and Content of Research-Oriented Activities of the World Employment Programme, Geneva 1972.

Little, I. M. D. and Mirrless, J. D., "Project Appraisal and Planning for Developing Countries", Basic Books Press, 1974.

"Multinational Corporations and Adaptive Research for Developing Countries", in Appropriate Technologies for International Development: Preliminary Survey of Research Activities, Washington, DC, September 1972.

Nelson, Richard R., "Less Developed Countries - Technology Transfer and Adaptation: The Role of the Indigenous Science Community", Economic Development and Cultural Change, Vol. 23, 61-78, 1 October 1974.

Penrose, Edith, "International Patenting and the Less Developed Countries", Economic Journal, 83, 331, 768-786, 1973.

Ranis, G., "Some Observations on the Economic Framework for Optimum LDC Utilization of Technology", in Economic Growth Centre Discussion Paper No. 152, Yale University.

Ravetz, J. R., Scientific Knowledge and Its Social Problems, Oxford University Press, 1971.

Rodriques, Carlos Alfredo, "Trade in Technological Knowledge and the National Advantage", Journal of Political Economy, 83, 1, 121-135, 1975.

Rosenberg, Nathan, "Factors Affecting the Diffusion of Technology", in Exploration in Economic History. Autumn 1972, Vol. 10, No. 1.

Ruais, G., "Source Observations on the Economic Framework for Opt. CPC Utilization of Technology", Economic Growth Centre, Paper 152, Yale University, 1973.

Schweitzer, Glenn, "Towards a Methodology for Assessing the Impact of Technology in Developing Countries", in Technology and Economics in International Development, AID, Washington, DC, May 1972.

Stewart, Frances, "Technology and Employment in LDCs", World Development, Vol. 2, 3, 17-46, March 1974.

Streeten, Paul, "Costs and Benefits of Multinational Enterprises in Less Developed Countries", in John H. Dunning, ed., The Multinational Enterprise, George Allen and Unwin, London, 1971.

The World Bank, "The Revision of the International Patent System: Legal Considerations for a Third World Position", World Development, 4, 2, 85-102, 1976.

The World Bank, The Role of Multinational Corporations in the Less Developed Countries' Trade in Technology, World Development, Vol. 3, 4, 161-189, April 1975.

UNCTAD, "Technological Dependence: Its Nature, Consequences, and Policy Implications", TD/190, prepared for Fourth Session, Nairobi, May 3, 1976.

UNESCO, A Methodology for Planning Technological Development, Report by Arthur D. Little, Inc. and Hetrick Associates, Inc., Paris, September 1970.

UNIDO, The Transfer of Technology in an Integrate Programme of Industrial Development, IPPD 53, Vienna, 15 October 1971.

United Nations, World Plan of Action for the Application of Science and Technology to Development, New York, 1971.

USAID, "Technology and Economics in International Development", Report of Seminar, Washington, DC, 1977, USAID Office of Science and Technology.

USAID, Technology and Economics in International Development, Report of the Seminar held in Washington, DC on 23 May 1972, Office of Science and Technology (TA/OST 72-9).

Vaitsos, Constantine V., "Foreign Investment and Productive Knowledge", in Guy F. Erb and Valeriana Kallab (eds.), Beyond Dependency: The Developing World Speaks Out, Overseas Development Council, Washington, DC, 75-90, 1975.

Valluri, S. R., Mobilization of National Resources and Planning of Industrial Research and Development, UNIDO.

Wionczek, Miguel, "Notes on Technology Transfer Through Multinational Enterprises in Latin America", Development and Change, Vol. 7, 135-155, 1976.

Appendix 6A - Basic Concepts of Game Theory or Strategic Competitive Decisionmaking

The Theory of Games is a series of methods applicable to systems in which two or more decision makers are in competition, and can be effectively employed in the analysis of licensee/licensor strategies. Such decision makers, who are not necessarily single persons, are called players, and we will designate players by p_1, p_2,....p_n (where n is the number of players). Players are concerned with objectives and to achieve their objective will adopt a "strategy". Strategies are the different decision alternatives of each player. Alternatives in turn may have numerous factors. If r is the number of factors and if the ith factor has m_i possible attributes, then the total number of alternatives is M = Product of all attributes. If the total number of alternatives is finite, the game is a "Finite Game"; otherwise, it is called a game of infinitely many strategies.

The outcome of a game is called the payoff. Payoffs are normally represented in matrix form (Payoff Matrix) which represents the result of the game from the point of view of the player under consideration.

A general Payoff Matrix is shown below:

P_2

P_1		B_1	B_2	....	B_n
	A_1	a_{11}	a_{12}		a_{1n}
	A_2	A_{21}			
	⋮				
	A_m	a_{m1}			a_{mn}

where A_1 are the alternative strategies of P_1, and B_1 are the alternative strategies of P_2 while a_{ij} are the payoffs for either P_1 or P_2. Obviously, if (a_{ij}) is the payoff for P_1, then $(-a_{ij})$ is the payoff for P_2.

A Zero-Sum game is the name for a game where the total payoff is zero or

where

$$(\Sigma a_{ij})_{P_1} = \text{Payoff of Player 1}$$

and

$$(\Sigma a_{ij})_{P_2} = \text{Payoff of Player 2.}$$

Non-zero-sum or constant payoff games on the other side result in a residue or input which is the difference in the payoffs of the players. In any game a player can adopt a <u>pure</u> or a <u>mixed strategy</u>. In the first he will choose the same alternative each game or each time a decision must be made, while in the second he will alternate his strategies. If P_1, for instance, plays a game N times and chooses decision alternative A_1, N_1 times and alternative A_2, N_2 times, then the frequency of decision A_1 is said to be

$$\frac{N_1}{N} = x_1$$

and A_2 is

$$\frac{N_2}{N} = x_2$$

where $x_1 \geq 0$, i = 1, 2,.....n and

$$\sum_{i \pm 1}^{m} x_i = 1$$

Similarly player P_2 chooses frequencies y_1, y_2....y_n where $y_j \geq 0$, j = 1, 2, ... n, and

$$\sum_{j=1}^{n} y_j = 1$$

We can describe a mixed strategy by vectors for:

P_1 by $X = (x_1, x_2, \ldots\ldots x_m)$

P_2 by $Y = (y_1, y_2, \ldots\ldots y_n)$

Now if P_1 chooses a pure strategy X_p, then X_p = (0, 0, 1, 0....0) and P_2's pure strategy might be Y_p = (0, 1, 0, 0....0), for example, given P_1 always chooses his third and Player 2 always his second decision alternative.

In order to judge the fairness of a game or our "chances" in a venture, we must develop an expression for the "Expected Payoff". If P_1 uses strategy X and P_2 uses strategy Y, then payoff a_{12} occurs with frequency x_1y_2; therefore

$$E(X, Y) = a_{11}x_1y_1 + a_{12}x_1y_2 + \ldots + a_{mn}x_my_n$$

$$= \sum_{i=1}^{m} a_{i1}x_iy_1 + a_{i2}x_iy_2 + \ldots + a_{in}x_iy_n$$

$$= \sum_{i=1}^{m} \sum_{j=1}^{n} a_{ij}x_iy_j$$

This is the amount P_1 gets and P_2, therefore, expects an amount -E(X,Y). The <u>Solution</u> of a game is the result of applying an <u>optimal strategy</u>. If P_1 uses $X_o = (x_{10}, x_{20}, \ldots\ldots x_{mo})$ which makes $E(X_o, Y)$ as large as possible, this is the solution for P_1. Similarly, $Y_o = (y_{10}, y_{20}, \ldots\ldots y_{no})$ is the optimal strategy for P_2.

If for any other relation of X and Y, we have

$$E(X, Y_o) \leq E(X_o, Y_o) \leq E(X_o, Y);$$

then $E(X_o, Y_o) = v_o =$ Value of the Game.

For the analysis of strategies, certain terms are found useful:

$\min_j a_{ij}$ = smallest term in row i in matrix a_{ij}

$\max_i a_{ij}$ = largest term in column j in matrix a_{ij}

$\max \min_j a_{ij}$ = largest element of terms min a_{ij}

$\min_j \max_i a_{ij}$ = smallest element of terms max a_{ij}

Analogous to the above, we can define functions

$$\min_y E(X,Y), \max_x E(X,Y), \max_x \min_y E(X,Y) \text{ and } \min_y \max_x E(X,Y)$$

The value of a game (v_o) is larger than the maximum that player P_1 can ensure of winning, and it is smaller than the min max that player P_2 can ensure of losing, or

$$\max_i \min_j a_{ij} \leq v_0 \leq \min_j \max_i a_{ij}$$

Consequently, a rectangular game in which

$$\max_i \min_j a_{ij} = \min_j \max_i a_{ij}$$

has a solution where optimal strategies are pure strategies and

$$E(X_0, Y_0) = \max_i \min_j a_{ij} = \min_j \max_i a_{ij}$$

Games like that will have a saddle point which is a point which is at the same time the lowest in the row and the highest in the column (or vice versa).

It can be shown that in a game in which the max min = min max = $v_o = a_{i_o j_o}$ or the value of the saddle point, this is the value of the game.

Payoff Matrix for a Game with a Saddle Point

		P_2					
Alternative		B_1	B_2	B_3	B_4	B_5	$\min a_{ij}$
P_1	A_1	-3	4	-2	0	-1	-3
	A_2	8	5	3	4	4	3
	A_3	3	1	2	1	2	1
	A_4	1	0	-1	2	0	-1
	$\max a_{ij}$	8	5	3	4	4	3

$$= \max_i \min_j a_{ij}$$

$$= \min_j \max_i a_{ij}$$

Therefore, value of this game = $v_o = E(X_o, Y_o)$, and the optimal strategy is $X_o = (0, 1, 0, 0)$ and $Y_o = (0, 0, 1, 0, 0)$ as a_{23} is the saddle point.

Next, let us consider Games without saddle points and starting with a simple 2 x 2 game; i.e., a game with two (2) alternatives per player.

Payoff Matrix

		P_2	
		B_1	B_2
P_1	A_1	a_{11}	a_{12}
	A_2	a_{21}	a_{22}

If the game has no saddle point, then $a_{11} > a_{12}$, $a_{11} > a_{21}$, $a_{22} > a_{12}$, $a_{22} > a_{21}$; the general solution of the game is $X_o = (x_o, 1\text{-}x_o)$ and $Y_o = (y_o, 1\text{-}y_o)$ where $0 \leq x_o \leq 1$, and $0 \leq y_o \leq 1$. Before going on, two theorems concerning rectangular games must be introduced.

Theorem 1

Every rectangular game has a value, and a player in a rectangular game has an optimal strategy.

Theorem 2

v^*, X^*, and Y^* are "value" and "optimal" strategies of players P_1 and P_2 only if for every pure strategy X_o of P_1 and Y_o of P_2, the following holds:

$$E(X_p, Y^*) \geq v^* \qquad E(X^*, Y_p) \geq v^*$$

It is important to notice that Theorem 1 does not imply that a rectangular game has a unique solution, while Theorem 2 is only a means of checking the proposed solution.

Returning now to our 2 x 2 game and using Theorem 2, we obtain:

$$a_{11}y_o + a_{12}(1-y_o) \leq v_o \qquad \text{for } A_1$$

$$a_{21}y_o + a_{22}(1-y_o) \leq v_o \qquad \text{for } A_2$$

$$a_{11}x_o + a_{21}(1-x_o) \geq v_o \qquad \text{for } B_1$$

$$a_{12}x_o + a_{22}(1-x_o) \geq v_o \qquad \text{for } B_2$$

It can be shown that these four inequalities have a solution only if the equalities hold. Therefore

$$v_o = [(a_{11}a_{22}) - (a_{12}a_{21})] / [(a_{12} + a_{22}) - (a_{12} + a_{21})]$$

$$X_o = (a_{22} - a_{21}) / [(a_{11} + a_{22}) - (a_{12} + a_{21})]$$

$$Y_o = (a_{22} - a_{12}) / [(a_{11} + a_{22}) - (a_{12} + a_{21})]$$

and Optimal Strategy for P_1

$$Y_o = [\frac{a_{22} - a_{12}}{(a_{11} + a_{22}) - (a_{12} + a_{21})} - \frac{a_{11} - a_{12}}{(a_{11} + a_{22}) - (a_{12} + a_{21})}$$

and for P_2

$$X_o = [\frac{a_{22} - a_{12}}{(a_{11} + a_{22}) - (a_{12} + a_{21})} - \frac{a_{11} - a_{12}}{(a_{11} + a_{22}) - (a_{12} + a_{21})}$$

Solution of a 2 x n or m x 2 game

Games where one player has only two alternatives can also be solved graphically, and the method will be presented here.

P_2

P_1		B_1	B_2	B_3	B_4
	A_1	11	-2	-4	4
	A_2	-4	3	6	-1

Payoff Matrix

If x* is the same frequency used by P_1, then

$$11x^* - 4(1-x^*) \geq v^*$$

$$-2x^* + 3(1-x^*) \geq v^*$$

$$x^* \geq 0$$

$$-4x^* + 6(1-x^*) \geq v^*$$

$$4x^* - 1(1-x^*) \geq v^*$$

Solution of m x n Games

These methods are all rather involved and will only be described briefly. Before attempting to solve a m x n game, it is always useful to look for a saddle point, and if such is not found, then for the occurrence of "Dominance".

Dominance is the term given to an Alternative A_i which is larger than another alternative A_i^1. We may then disregard A_i^1 and reduce the number of alternatives to a 2 x n or m x 2 matrix.

Linear Programming Solutions

Every game problem can be transformed into a linear programming (LP) problem. One of the essentials in an L.P. problem is that payoffs must be non-negative, and we therefore have to add a constant (if required) to all terms in our payoff matrix. If we take, as an example, our 2 x 4 matrix and add a constant 5, we obtain:

P_2

P_1		B_1	B_2	B_3	B_4
	A_1	16	3	1	9
	A_2	1	8	11	4

The value of the game is now

$$g_o = v_o + 5 \qquad g_o > 0$$

(It can be shown that optimum strategy is not affected by the addition of the constants.) Let $X^* = (x_1^*, x_2^*)$, $x_2^* = 1 - x_1^*$; then

$$16x_1^* + x_2^* \geq g^*$$
$$3x_1^* + 8x_2^* \geq g^*$$
$$x_1^* + 11x_2^* \geq g^*$$
$$9x_1^* + 4x_2^* \geq g^*$$

Dividing by g*, we obtain

$$\begin{aligned}
16\, x_1^*/g^* &+ x_2^*/g^* \geq 1 \\
3\, x_1^*/g^* &+ 8\, x_2^*/g^* \geq 1 \\
x_1^*/g^* &+ 11\, x_2^*/g^* \geq 1 \\
9\, x_1^*/g^* &+ x_2^*/g^* \geq 1 \\
x_1^*/g^* &+ x_2^*/g^* \geq 1/g^*
\end{aligned}$$

If we put

$$\begin{aligned}
s_1 &= x_1^*/g^* \\
s_2 &= 1/10 \\
n &= 1/g^*
\end{aligned}$$

then

$$\begin{aligned}
16s_1 &+ s_2 \geq 1 \\
3s_1 &+ 8s_2 \geq 1 \\
s_1 &+ 11s_2 \geq 1 \\
9s_1 &+ 4s_2 \geq 1 \\
n &= s_1 + s_2
\end{aligned}$$

P_1 is interested in maximizing g_o; i.e., g^* or in minimizing $1/g^* = n$. Using a graphical or the simplex method, we obtain $s_1 = 1/15$, $s_2 = 1/10$, $n = 1/6$, $g_o = 6$, $x_1 = 2/5$, $x_2 = 3/5$, $x_o = (2/5. 3/5)$, $v_o = 6\text{-}5 = 1$.

Continuous games are normally solved graphically, and apply to games where players can choose any number between 0 and 1. Continuous games always have a solution which can be obtained by plotting or assuming small incremental changes until a solution is found.

Non-zero-sum games (closed bidding) are often solved by introducing a fictitious player P_3 (for instance) with a single alternative. P_3 is then considered not an active participant in coalition, and the game is transferred into a zero-sum game.

Let us next consider a simple zero-sum n person game. One approach to the solution of such a game is to assume that various players are in coalition. If in a four person zero-sum game, for example, three players A, B, and C form a coalition against D, and the value of the game for the coalition is v, then the value of the game for D is obviously equal to -v. In such a four person game we have seven such values or characteristics for the game.

Let us now consider a three person zero-sum game in which each player has two alternative strategies. Therefore, player A can use his strategy A_1 or A_2, player B his strategy B_1 or B_2 and player C his strategy C_1 or C_2.

The payoff matrix in such a n person game shows on each line one possible set of selection alternative strategies by the players and the number of rows in the payoff matrix is equal to the number of combinations of player strategies. In a three person game with two alternative strategies per player, we have $2^3 = 8$ different rows at outcomes. Considering the payoff of the individual player, the sum of the payoffs in each row of a n person zero sum game must add to zero as shown in the payoff matrix.

Player			Payoff		
A	B	C	A	B	C
A_1	B_1	C_1	2	1	-3
A_1	B_1	C_2	-1	1	0
A_1	B_2	C_1	-1	-2	3
A_1	B_2	C_2	0	2	-2
A_2	B_1	C_1	3	-2	-1
A_2	B_1	C_2	-2	0	2
A_2	B_2	C_1	0	-1	1
A_2	B_2	C_2	-1	1	0

In order to get a solution domain let us next assume that players form coalitions which transform the above game into three two person games in which

A plays a coalition of BC

or

B plays a coalition of AC

or

C plays a coalition of AB.

We furthermore assume that these are perfect coalitions in which parties agree on an optimum joint action.

Considering A versus BC, first we obtain the following two person payoff matrix for player A.

		COALITION				
		$B_1\,C_1$	$B_1\,C_2$	$B_2\,C_1$	$B_2\,C_2$	Min
	A_1	2	-1	-1	0	-1
A	A	3	-2	0	-1	-2
	Max	3	-1	0	0	-1

Therefore, this game has a saddle point as the min max = max min = value of game = -1. Therefore, v(A) = -1 and v(BC) = +1.

Looking next at B playing the coalition of AC.

		COALITION			
		$A_1\,C_1$	$A_1\,C_2$	$A_2\,C_1$	$A_2\,C_2$
B	B_1	1	1	-2	0
	B_2	-2	2	-1	1

Here, we note that coalition strategy A_1C_1 dominates strategy A_1C_2, and similarly strategy A_2C_1 dominates strategy A_2C_2. As a result, we can replace the above game by a simple two person 2 by 2 game.

	A_1C_1	A_2C_1
B_1	1	-2
B_2	-2	-1

Assuming the frequency with which B plays his strategy B_1 is x and that of his strategy B_2 is (1-x), the following inequalities apply.

$$x - 2(1-x) \leq v(B)$$

$$-2x - (1-x) \leq v(B)$$

For these inequalities to hold

Finally, considering a game of C versus a coalition of A and b

	A_1 B_1	A_1 B_2	A_2 B_1	A_2 B2
C_1	-3	3	-1	1
C_2	0	-2	2	0

The value of this game can readily be shown to be equal to v(C) = -0.75. We, therefore, have a set of solutions for this game with each player playing against a coalition of his opponents. This gives us a bound because these solutions are the worst a player can do.

To find the solution of the game we must now develop a set of imputations which are 'results of the game' for each player in such a way that no player wins less (or losses more) than when playing against the coalition of opponents and in which the sum of the outcome for all three players is zero.

The bound of this game is as we established

$$x_1 \geq -1,\ x_2 \geq -1.25,\ x_3 \geq -0.75$$

If we define an imputation by a vector $X = [x_1, x_2, x_3]$ in which $x_1 = 0$ and the above constraints hold, then a set of imputation is:

-1	-1	+2
-1	-1.25	+2.25
-1	+1.75	-0.75
,	,	,
,	,	,

Considering a set of imputations as above we now consider the set of possible solutions represented by the imputations such that

1. No imputation in the set dominates another imputation in the set.

2. Any other imputation is dominated by one of the imputations in the set.

If we continue and consider the four imputations:

-0.75	0.25	0.50	(1)
2.00	-1.25	-0.75	(2)
0.30	-0.80	0.50	(3)
-0.15	0.85	-0.70	(4)

but not:

0.50	0.50	-1.00

it can be seen that

B and C prefer (1) to (2) or (4) to (2)

A and B (4) to (1)

Therefore, we can say that

(4) dominates (1) with respect to A and B

and finally

K = [0.50, 0.25, -0.75]

L = [0.50, -1.25, 0.75]

M = [-1.00, 0.25, 0.75]

For example, if we take a simpler set of characteristic functions

v(A, B) = -1	v(C) = 1
v(B, C) = 2	v(A) = -2
v(A, C) = 3	v(B) = -3

then

$x_1 \geq -2$; $x_2 \geq -3$; and $x_3 \geq 1$

Using constraints $x_i \geq v_i$, and $\Sigma x_i = 0$, we obtain

Imputation No.	Imputation			Comment
1	-1	-1	2	
2	0	-2	2	
3	-1	-2	3	
4	2	-3	1	
5	0	-1	1	
6	-2	1	1	
7	0	-3	3	
8	-2	-1	3	

Let us consider a sample n person non-zero sum game. The next common solution approach is to transform the game into a (n+1) zero sum game by introducing a fictitious paly n+1. This player is assumed not to be free to join a coalition. In other words, we consider the fictitious player only from the point of view of maximum possible loss.

Let us assume that $g_1, g_2 .... g_n$ are the values of the game for players 1 to n playing against a coalition and G = g(1....n) is the value of the game of all players playing against the fictitious player. Then

$$G = g(1....n) = -g_{n+1}$$

If again

$$x_i \geq g_i \text{ and } \sum_{i=1}^{n} x_i = G$$

Taking a simple example

Player		Payoff	
A	B	A	B
A_1	B_1	3	-1
A_1	B_2	-1	1
A_2	B_1	1	0
A_2	B_2	-2	-1

Player A

	B_1	B_2
A_1	3	-1
A_2	1	-2

g(A) = -1

Player B

	A_1	A_2
B_1	-1	0
B_2	1	-1

g(B) = -1/3

Player (n+1) = C

	A_1B_1	A_1B_2	A_2B_1	A_2B_2
C	-2	0	-1	+3

g(C) = -2

Therefore

$g_1 = -1 \qquad g_2 = -1/3 \qquad g_{n+1} = -2$

and

$G = -g_{n+1} = -g_3 = 2$

but in our imputations

The result is therefore: [-1. 3]. [2 1/3, -1/3], [0, 2], etc.

Game theory provides a useful tool for the development of an effective strategy in protection or acquisition of technology. Estimates of expected payoffs by competing parties, say the inventor and the investor or technology acquirer, often vary widely and seldom conform to a zero sum situation. Inventors usually have a number of alternative strategies from outright sale of all rights to delay of sale of any rights or results until complete development of the invention. Technology acquirers similarly may simply want access to a new technology at a lower cost and less time, that it would take the acquirer to obtain the technology otherwise, if at all possible.

Payoff matrices usually represent estimates of expected payoffs over the life of the invention, given a set of strategies are adopted by the inventor and the acquirer. On the other hand, many real world situations require consideration of the effects of strategy opportunities of third parties, such as inventors or developers of competing technologies, other acquirers, and potential users. Development of larger multi-person competitive technology development and acquisition models or games goes beyond the scope of this book, as does the inclusion of conditional stochastic continuous technology acquisition games.

7. Technology Pricing and Economic Evaluation

The price of technology depends to a large extent on its usefulness, effectiveness, desirability, competitiveness, and acceptability. The price may also be affected by the publicity the technology has received, and the strategy of the owner of the technology, which will usually be influenced by expected developments impacting on the position of the technology. Developments which must be considered are:

a. the characteristics, extent, and development of the demand for the technology

This in turn will be affected by

b. emergency of competing technologies or improvements in existing technologies, and
c. the resulting competitive position, both short and longer term, of the new technology.

Other issues are concerned with financial and marketing strategies of the owner, and constraints or restrictions imposed on the owner of the technology by government, trade associations, international agencies, or others.

The price of the technology will affect its use or diffusion, as well as the development of competing technology. If it is priced too high, not only will fewer buy it, but the high price will also encourage potential users to upgrade their existing technology. Competing technology may also be developed by other developers who are not potential users. If the new technology represents a radical change which offers new product, process, or service technology with unique and desirable characteristics which offer major advances and advantages over those of existing technology, a larger price can usually be demanded by the owner for the use of the technology.

Radical change in technology offer radical improvements in competition. Such radical changes in technology, often the result of radical innovation which is innovation undertaken at great speed and large risks with the goal of leapfrogging existing technology and technology under development, cause major shocks in industry by making existing, often young technology obsolete long before its reaches maturity or approaches the end of its operating life. Radical innovation of technology is usually driven by crisis, unique opportunities, or threats. Such crises may be caused by radical changes in demand, large deviation in economic factors, such as labor costs, and most importantly emergence of radically different and/or improved technology.

The response to radical change in technology may be to broaden the use of existing technology or search for new combinations or coalitions of technology. Often new opportunities can be developed for existing technology in this manner, and the impact of the new technology reduced or even neutralized.

Often the response to radical innovation in technology is to delay the existence of a threat or opportunity. As a result, little or no defensive action is taken by both competitors and potential users. They may in fact act against their own best interest, or react by emphasizing the old technological approach. It is important for owners of new technology to understand the dynamics of technological change and particularly that of radical technological innovation. Only then will they be able to develop and use an effective technology pricing and diffusion strategy.

Changes in technology invariably introduce pressures on the demand for new process, product, or service technology. This is sometimes the result of the change in production or service costs or other measures of performance of the new technology. In a process technology, the effect can usually be noticed in the change of the slope of the learning curve as new technology replaces existing technology. Considering macro effects of technological change, the rate of diffusion and the impact of the new technology depends on many factors such as:

1. state of technology in terms of status of innovation or application development stage;
2. ownership and accessibility of technology - Is it closely held and guarded? Is the patent or rights holding monopolistic?
3. market for technology - Is it broad? Do owner(s) of new technology have monopolistic dominant, minor, or no market share? Do owners use a competing technology, if any, to satisfy their market component?
4. competition - Is the competition, if any, for the market dispersed, powerful, dominating? Do competitors produce, use, or have access to substitute technology? How does substitute technology compare with the new technology in terms of cost, quality, performance, etc.?
5. pricing of new technology and terms under which it becomes available,
6. effectiveness of technology transfer and absorption capability.

In case the new technology is not widely diffused among suppliers or users but introduced monopolistically, and has major cost and/or performance advantages over existing and competing technologies, if any, then given the demand for the new technology is elastic, the monopolistic owner will assume a profit maximization strategy and price the technology at a level consistent with the demand/price at the output level where marginal costs are equal to marginal revenue as shown in Figure 7.1, and produce a quantity Q_o at a Cost C_c. The price charged would then be P_o and the resulting profit $(P_o-C_c)Q_o$.

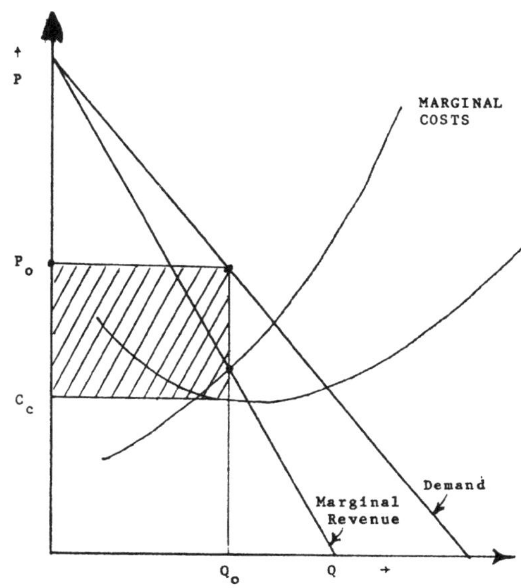

Figure 7.1. Monopolistic Pricing of New Technology.

If the new technology is not monopolistically controlled, then its introduction will not only affect its costs and performance, but also its price, as users of the new technology adjust their price to change their market share and total profit. Consider a simple experience curve situation (Figure 7.2), where an existing technology achieves cost improvements as output increases. While the price for the output - after introducing the new technology declines and the slope of the price curve - increases, the profit or difference between price and cost is gradually reduced, with a resulting narrowing of the profit margin. Given a competing new technology is available, competitors will increasingly consider its adoption, even if entry costs are high and the old technology has not yet reached the end of its economic life. The trigger to adopt the new technology is usually released when a competitor drastically changes the price trend as shown with a resulting change in the slope of the price curve. Some, and particularly the larger producers, among the competitors will then search for a technology which promises a cost curve with a slope exceeding that of the new price curve and will adopt this new technology, thereby enhancing the rate of its diffusion and the resulting rate of transition from the old to the new technology cost performance curve as shown. Usually when a price break (change in price curve slope) occurs, the rate of diffusion or introduction of new technology will accelerate. If the old technology is gradually replaced by the new technology starting at Q_1, a cost curve could be achieved which broadens the profit margin again as shown in Figure 7.2.

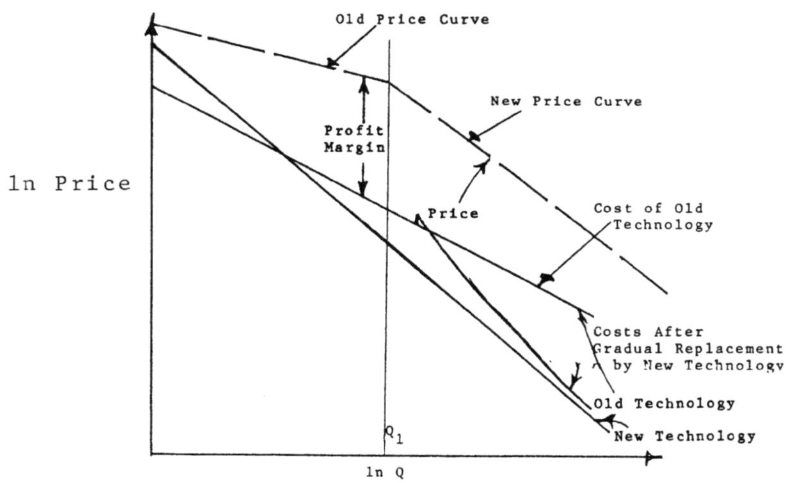

Figure 7.2. Price Change and Introduction of New Technology.

In economic terms, this can be considered a rapid change in the long-run average cost curves with a discontinuous long-term average cost curve and a new long-run marginal cost curve as shown in Figure 7.3. Obviously the degree of discontinuity will be a function of the rate of introduction of the new technology. If this rate is very gradual, then instead of a change in slope of the productivity (cost or performance) curve, a gradual curving would occur.

Given the new technology user is a monopolist and tries to operate at his optimum output when in long-run equilibrium, then the long-run average cost curve (LAC) must cross the marginal revenue (MR) line at its lowest point (Figure 7.4). If the MR curve crosses the LAC curve to the right of the lowest point on the LAC curve, then the monopolist overutilizes the new technology and uses a larger than optimum scale of in long-run equilibrium. Similarly if the MR curve crosses the LAC curve to the left of the lowest point on the LAC curve, the monopolist underutilizes the technology and uses a smaller than optimum scale in long-run equilibrium. It should also be noted that the MR curve would change with a change in price and in fact with rapid learning and/or change in technology, a continuous change in the price and MR curve must usually be assumed.

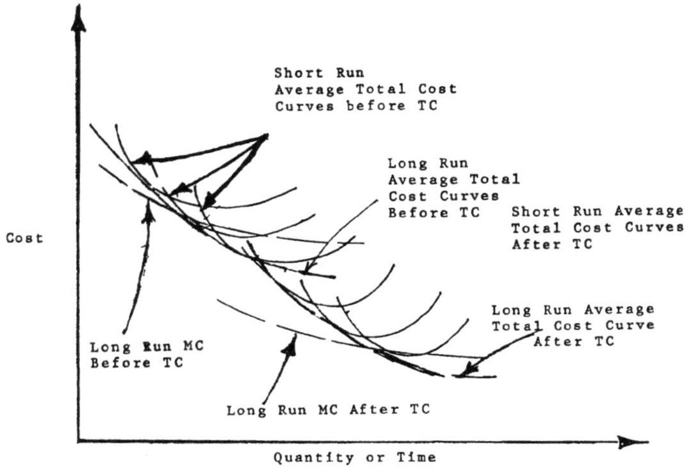

Figure 7.3. Break in Long Run Average Total Cost Resulting from Technology Chance (TC).

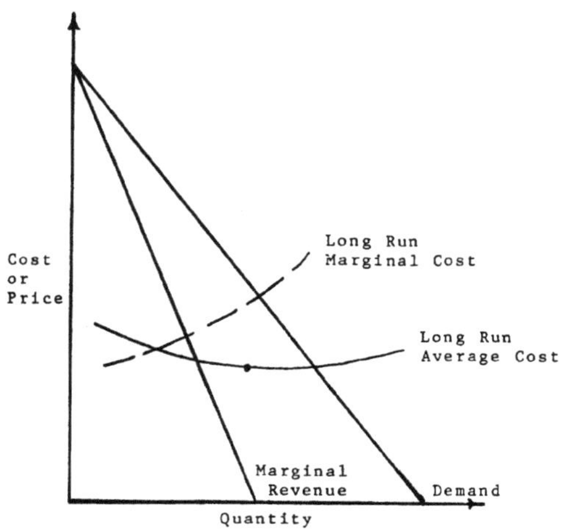

Figure 7.4. New Technology Utilization by Monopolist.

In his recent work Spence [1] evaluated the effect of competition under conditions of learning. He showed that when learning occurs, part of a firm's marginal costs should be regarded as an investment which reduces future costs of production or improvements in performance. If there is no competition, a profit maximizing oligopolistic firm will reduce its prices always more slowly than costs whereby it causes short run margins to widen over time.

When competition exists or could potentially emerge though, a firm will reduce its prices as learning proceeds and try to maintain profit margins. The rate of price reduction in relation to cost reduction depends on:

1. the degree and power of competition;
2. availability and diffusion of new technology with a higher learning curve slope and starting costs not much above current costs of production; and,
3. the current level of the profit margin.

Lieberman [2] modeled price developments in the chemical processing industry to determine the effect of learning on pricing and found that prices "declined in close conformity with the learning curve model". He also found a marked difference in the short-run price behavior, between high and low concentration markets.

7.1 PRICING NEW TECHNOLOGY - IMPACT OF DIFFUSION AND MARKETING

The price for a new technology is affected by its state of development, the degree of protection provided or maintained, its usefulness and the potential demand for it, the demand and elasticity, as well as regulation imposed on its use by government, and most importantly the existence and role of competing technologies. Until technology is generally diffused the price-cost gap will usually increase, but as diffusion of technology accelerates, which happens often in the later stages of innovation, the price-cost differentials will usually start to decline until the gap narrows to a small fraction, when diffusion becomes universal and the technology reaches maturity as shown in Figure 7.5. This theoretical development contradicts marginal cost pricing and is largely based on a concept of opportunity or value pricing.

In many cases diffusion of new technology is delayed to well into the innovation stage. This, as explained before, would happen if patenting and delayed or restrictive licensing is used by the technology owner and/or innovator. Government regulation may have a price stabilizing effect, under which price is maintained independent of changes in costs, though this is often introduced to compensate the technology owner/innovator for up-front losses.

Similarly marketing can have a price maintenance or price depressant effect, depending on the sales objective, degree of competition, and demand elasticity. Demand elasticity, which is usually a major determinant of marketing strategy, also affects both short and long run pricing tactics. A technology owner may, for example, find it profitable to speed up output periodically with corresponding price reductions to take advantage of demand elasticity, and simultaneously undercut competition to increase market share.

Similarly, the price of the technology will obviously affect the rate of diffusion at a particular point in the innovation stage. Innovations usually result in new, and often better, products or processes whose diffusion would result in greater factor productivity, through increase in productivity and/or capacity, which in turn generally causes a lowering of the unit costs of production. In the short run we would therefore expect increases in price to provide an increase in marginal revenue as price declines at a lower rate than cost. When diffusion becomes more pronounced and competitive pressures result in a sharp decline in price, the rate of production is usually accelerated as well causing a more rapid decline in costs. This is seldom sufficient to maintain the profit per unit output but total profit often increases as a result of larger scale production.

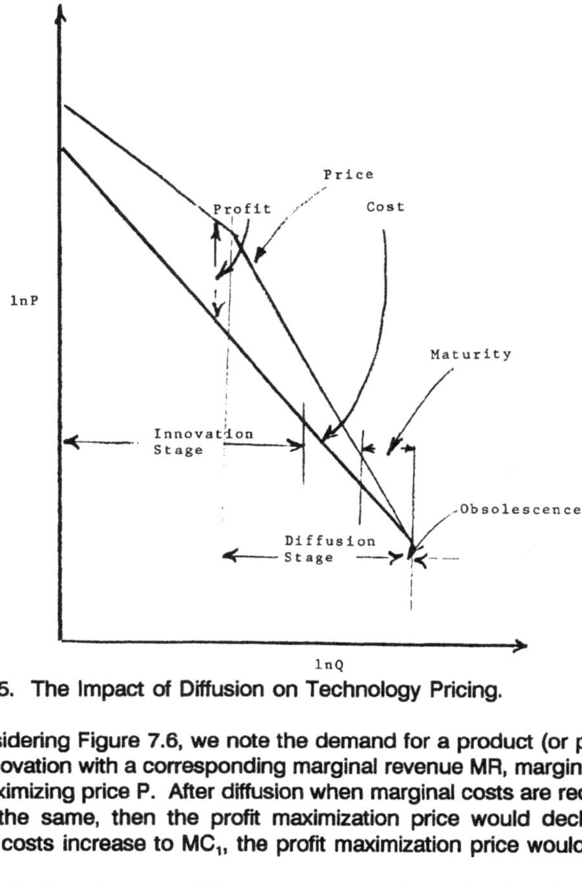

Figure 7.5. The Impact of Diffusion on Technology Pricing.

Considering Figure 7.6, we note the demand for a product (or process) D before diffusion of the innovation with a corresponding marginal revenue MR, marginal costs, MC, and resulting profit maximizing price P. After diffusion when marginal costs are reduced to MC_2 and demand remains the same, then the profit maximization price would decline to P_2. (Conversely if marginal costs increase to MC_1, the profit maximization price would increase to P_1.)

In reality though, competition may cause a change in the price and thereby the marginal revenues to MR_1 and MR_2 as shown which would cause a new price adjustment. In fact, during increasing diffusion of an innovation, these price reductions are made frequently with a resulting change in marginal revenues and marginal costs as production or output is adjusted to the new reality of cost and price. The adjustment though often causes new changes. There is therefore a feedback process which becomes even more complex if the demand curve changes as well. A change in the demand curve is usually more prevalent in product innovation. Improvements in a product may not only shift the demand curve but affect its elasticity as well.

The main issue in an economic evaluation of the effect of pricing is the degree of control the innovator maintains and how the control is used. Another important factor is obviously the existence of competing products and processes which, though different, may sell in or affect the same market, as the product or process technology of interest.

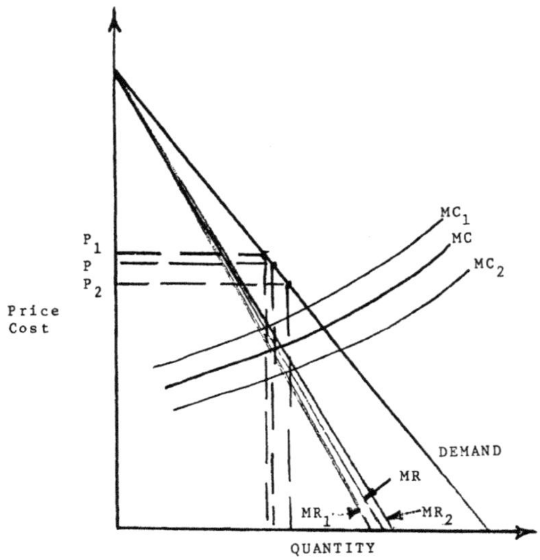

Figure 7.6. Economic Effects of Diffusion of Innovation.

7.2 TECHNOLOGY PRICING STRATEGIES

The pricing of technology is affected by endogenous and exogenous factors. Endogenous factors include:

1. Innovator or Owner Personality
   Corporation, individual, government agency, etc.

2. Innovator or Owner Objectives
   Profit, market share, quality, monopoly control, competitor elimination, knowledge expansion, cost reduction, etc.

3. Innovator or Owner Control
   Invention ownership and rights, state of innovation, market control, marketing ability, competing technologies, internal cost control, quality control, etc.

4. Innovator or Owner Capability
   Financial resources, capacity, etc.

Endogenous factors influencing pricing strategies can be summarized as:

1. Competition
   Competitors, competing products and processes, market characteristics, market control, and market demand, etc.

2. Regulation
   Trade, environmental, economic, quality, product, etc.

3. Market Characteristics
   Market composition, market developments, market sectors, market distribution, etc.

In addition, we must usually consider the political and social environment in which the technological change occurs.

Technology pricing strategies therefore are difficult to determine and require nimble and effective decisionmaking. Pricing of technology must be dynamic and conditional and should be affected by the above factors with due consideration to the ability of the innovator and/or technology owner/licensee to change the rate of output (capacity) and costs to respond to price changes and therefore marginal revenue and profit changes often induced exogenously.

Technology pricing strategies can be protective, assertive, responsive, or defensive. In each, the decisions will affect price, cost, rate of output, and change in capacity decisions. Furthermore the innovator may encourage or discourage the rate of technology diffusion and thereby external innovation of his technology. This type of decision will often be influenced by (1) the current and expected profit by the inventor or technology owner from the invention or innovation, (2) market conditions and the technology owners' market share, (3) competition, (4) expansion capability and financial resources of the technology owner, (5) inventor, innovator, or technology owner personality and objectives, and (6) control of the inventor, innovator, or technology owner over the technology.

7.2.1 *Technology Pricing Analysis.* The determination of a pricing strategy, and particularly a strategy for pricing the use (or output) of a new product, service, or process technology is difficult, as it depends on many exogenous and endogenous factors, as well as objectives and goals which drive the pricing strategy decisions.

Various macro and micro economic technology pricing models have been proposed. While the macro models deal mainly with the impact of technological change on macro (world, national or regional) benefits, many of the microeconomic models evaluate the effect of technology cost and price on the profitability of technology use. The difficulty with these models though is that they are usually static and assume prices to be invariant with quantity or rate of output produced during a period of time or cumulative output produced by the technology.

An interesting stochastic model for technology pricing was developed by Schwartz et al [7]. Suppose that the duration of the technology's life (t), from the present moment (moment when the planning period starts) until the moment t+z, when the technology is fully and finally preempted by another one is random. Let us also denote by F(t) the distribution of the random variable z.

Dividing the planning period into n intervals, we can express the probability Z_i that the entire process will stop at the end of the ith interval, given it has not stopped before, as

$$Z_i = \frac{F(t_i) - F(t_{i-1})}{1 - F(t_{i-1})}$$

where

$$F(t_i) = \sum_{k=2}^{i} (1-Z_1)(1-Z_2)\cdot\cdot\cdot(1-Z_{k-1})Z_k \qquad (7.1)$$

Let us assume that the demand in any given period follows a distribution denoted by $f_d(d_0)$, which incorporates the effect of sales of the relevant technology in previous years.

Suppose also that unit product cost is g and the price is C if supply is totally met by demand, and excess supply can be sold for a lower price P = aC, where o < a < I. The producer's profit, if we denote y as the production and d_0 the demand, is

$$C \min(d_0, y) + \alpha C\{y - \min(y, d_0)\} - gy \qquad (7.2)$$

and its expected value is:

$$\pi = (C-g)y - C(1-\alpha)\int_0^y (y-d_0) f_d(d_0) d(d_0) \qquad (7.3)$$

In the case of unconstrained demand (3) reduces to:

$$\pi = (C-g)y \qquad (7.4)$$

Maximization of (3) results in an optimum value for y (supposing that no additional constraints are imposed), which if substituted in (3) gives the mean of the maximum profits:

$$\pi_{opt} = (1-\alpha)C \int_0^{y_{opt}} d_0 \cdot f_d(d_0) d(d_0)$$

If we do not know $f_d(d_0)$ but only m = E(d) = mean demand and D, we can use a worst case analysis which leads to

$$\pi = \{(C-g)y - C(1-\alpha) \int_0^y (y-d_0) \cdot f_d(d_0) d(d_0)\} \qquad (7.5)$$

which becomes maximum for

$$y_{opt} = \begin{cases} m + D \cdot f(b) & \text{if } (1 + \frac{D}{m^2}) < 1 \\ 0 \text{ otherwise} \end{cases}$$

where

$$b = \frac{1}{1-\alpha}\left(\frac{g}{C} - \alpha\right)$$

$$f(b) = \frac{1-2b}{2\sqrt{b(1-b)}}$$

This value of y_{opt} results in

$$\pi_{opt} = \max[(C-g)m - \sqrt{D}\sqrt{(g-\alpha C)(C-g)},\ 0] \qquad (7.6)$$

We assume that when the seller chooses his technology trade strategy, he estimates m, D, or, if possible, $f_d(d_0)$. For the sake of simplicity, we suppose that in a span of m periods of time, the sale of current technology may occur no more than once in a period. We shall suppose that the market initially available for product sales consists of N units and measure a sale in these units, so that technology sale of a unit leads to market reduction of a unit.

Let g_i^S be the probability of seller being offered the transfer of (a) units of technology in the ith period after he has sold (s) units of the total N he had. Define variables

$$U_i^s(a) = \begin{cases} 1 & \text{if he accepts the other} \\ 0 & \text{otherwise} \end{cases} \tag{7.7}$$

The strategy $V = (U_i^s(a))$ may be optimized given price $p_i^s(a)$ and profits from product sales by a seller during the ith period.

The conditional mean profit of the seller in the ith time period that begins after technology worth s units has been sold, is given by:

$$V_i^s = \sum_{\alpha=1}^{N-S} g_i^s(a)[U_i^s(a)[P_i^s(a) + \pi_i^{s+a}] + (1-U_i^s(a)]\pi_i^s$$
$$+ [1 - \sum_{\alpha=1}^{N-S} g_i^s(a)]\pi_i^s, \quad i = 1,2\ldots n \tag{7.8}$$

The optimization problem is to choose $U_i^s(a)$ maximizing the sum of V_i^s's for all i due to the correspondent sequence $S=s(i)$.

Dynamic programming is used to choose $U_n^s(a)$ and all s and a, maximizing V_n^s, then, to choose $U_n^s(a)$ for all s and a maximizing the two period profit, etc. If we denote W_i^s as the maximal mean profit during periods i, i+1...n, given the state after (i-1) periods is s, the following recursion formula results:

$$W_i^s = \max\ (V_i^s + \sum_K q^K_{sk}\ W^K_{i+1})\ U_i^s(a) \tag{7.9}$$
$$,\ i = 1,\ 2,\ \ldots n-1$$

where q_{sk}^i are transition probabilities for the ith period. Both W_i^s and q_{sk}^i depend on $U_i^s(a)$. Substituting q_{sk}^i we obtain

$$W^s_i = \max\ (\sum_{a=1}^{N-S} g^2_i(a)[(U_i^s(a)[P^s_i(a) \tag{7.10}$$
$$+ \pi^{s+a}_i - \pi^s_i] + \pi^s_i]\ U^s_i(a)$$
$$+ \sum_{a=1}^{N-S} g^s_i(a)U^s_i(a)(1-Z^{s+a}_i)W^{s+a}_{i+1}$$
$$+ [1-\sum_{\alpha=1}^{N-S} g^s_i(a)V^s_i(a)][(1-Z^s_i)W^2_{i+1}]$$

This recursion formula is solved backwards.

Rationally thinking we can say that the prices when all $U_i^s(a) = 1$, should be greater than when some $U_i^s(a)$ are equal to zero,.

Denoting with (') the symbols refer to the case that all $U_i^s(a) = 1$ we have

$$p_i^{s'}(\alpha) \geq \pi_i^s - \pi_i^{s+a} - (1-Z_i^s)W_{i+1}^s - (1-Z_i^{s+a})W_{i+1}^{s+a} \tag{7.11}$$

The system of inequalities (11) can be solved recursively. If now we define boundary prices $P_i^s(a)$

$$\tilde{P}_i^{\,s}(\alpha) = \pi_i^{\,s} - \pi_i^{\,s+a} + (1-Z_i^{\,s})W_{i+1}^{\,s} - (1-Z_i^{\,s+a})W_{i+1}^{\,s+a} \qquad (7.12)$$

It turns out to be:

$$\tilde{P}_i^{\,s}(\alpha) = \pi_i^{\,s} - \pi_i^{\,s+a} + \sum_{k=i+1}^{N} (\pi_k^{\,s} \prod_{k=j=1}^{k-s} (1-Z_j^{\,s}) - \pi_k^{\,s+a} \prod_{j=1}^{k-s} (1-Z_j^{\,s+}$$

These boundary prices may be interpreted as the lower limits of the real prices. An interesting characteristic of the boundary prices is their independence of the probability $g_i^s(a)$. This may be explained as follows: the boundary prices were introduced by compensate the seller's losses in the product sales caused by his market reduction as a result of technology sale. If the moment i no future sales of technology may be predicted for certain, all losses caused by the decision to sell the technology at this moment must be compensated by $P_i^s(a)$. This is seen more clearly when the deadline of the price is fixed, that is every $Z_{i=0}$ for i<n. In this case

$$P_i^{\,s}(\alpha) = \sum_{k=1}^{n} (\pi_k^{\,s} - \pi_k^{\,s+a}) \qquad (7.13)$$

7.3 PRICING MATURE TECHNOLOGY-SUBSTITUTION EFFECTS

Pricing of mature technology or technology well advanced in the innovation process depends on the degree of control (or monopoly) maintained in that technology and the prospects of competing technology. It is common for mature technology to already be widely diffused and under increasing pressure by substituting technologies. As a result, few mature product or process technologies can affect the competitive environment.

Their principal advantage in case of process technologies is usually low fixed and marginal costs, with capital investments largely written off and productivity nearly at fully experienced levels. Additional marginal cost reductions, if at all achievable, are usually small and no longer dependent on large additions in cumulative output. Pricing must therefore take full advantage of the low fixed costs, but can no longer depend on major cost reductions from large increases in the rate of production or output.

Under these circumstances the substitution effect, which refers to both substitution of the process by new technology as well as substitution of output by different products manufactured by similar or different processes, and the obsolescence effect which causes large increases in unit costs (or loss of productivity) due to process efficiency loss, condition of process (wear out), difficulty in maintaining process, and more, must be considered in price determination and decisions affecting the extent of usage of the old technology. Most importantly the slope of the price curve, which in turn is affected by the existence of competing technologies, must be considered in such decisions.

7.4 COST/BENEFIT ANALYSIS OF LEARNING AND TECHNOLOGICAL CHANGE

The introduction of new technology requires capital, installation, run-in, and training expenditures, which become a fixed cost and can be distributed over future output or use. Unfortunately the ultimate cumulative output of a process or technology is usually not known at acquisition and therefore estimates of economic cumulative output quantities are usually used to determine the fixed cost component per unit of output. Variable costs are then assumed to benefit from the learning. If the technology or process is changed or its use discontinued before achievement of the economic cumulative output (or economic life), then fixed costs should either be increased over the cumulative output produced so far or the unamortized fixed costs must be added to the costs of the new process or technology acquired as a transaction cost to be amortized over the economic cumulative output of the new technology or process. Obviously the salvage or resale value of the old (replaced) process or technology should be deducted from the residual (unamortized) fixed costs. If the salvage or resale value is larger than this residual, then a negative net residual fixed cost would occur which should be redistributed over the cumulative output achieved before the discontinuation.

Another issue is the consideration of the cost of time in the allocation of fixed costs. The economic cumulative output from a process or technology can be produced at various rates of output until the economic life in terms of cumulative output is reached. The economic life of the process may therefore be shorter or longer. As fixed costs per unit output include both capital repayment (or amortization) and interest charges, they will vary with the time required to produce the economic cumulative output.

Similarly if the use of the technology is discontinued before the economic cumulative output is reached, fixed costs must be recalculated and residual fixed costs (including interest) reassigned, taking due account of the time required to reach the level of cumulative output at which the technological change was introduced. The benefits of technological change include such factors as:

Quantitative Benefits	Qualitative Benefits
Cost Reduction	Quality Improvement
Revenue or Profit Increase	Knowledge of Enhancement
Improvements in Market Share	Competitive Edge
Market Value	Reputation
Capacity Increase	Working Conditions

Similarly, costs of technological change can usually be listed as:

Quantitative Costs	Qualitative Costs
Capital Investment	Environmental Impact
Training Costs	Working Conditions
Redeployment of Workers	Organizational Costs
Installation and Testing Costs	Performance Risk
Run-In and Start-Up Costs	
Operating Costs	

Estimates of costs and benefits of a technological change are usually developed to support technological change decisions. Such estimates often include quantitative equivalents of qualitative costs and benefits.

Summary costs and benefits are next divided into fixed benefits and costs which may be independent of the rate of output but dependent on the expected time and total output the process or product technology is used and variable benefits and costs which are rate of output dependent. In other words, fixed costs and benefits are often a direct function of time, while variable costs are benefits of output. There are obviously products and processes whose life is a function of a cumulative output. For effective valuation though, both have to be discounted to present value terms. Considering process technology for example, an important consideration is therefore its economic life, which can be expressed both in terms of time and cumulative output. This often raises a problem, particularly when a process is said to have an economic life of say N cumulative units of output or T years whichever occurs earlier, where T is the assumed time when the technology is deemed to be obsolete and to no longer serve its assumed function. This means that some user may discard the process long before obsolescence, because it is worn out (has produced N units of output), while other users may discard it because it is obsolete even though it has not yet produced its designed cumulative output N. The difficulty arises from the fact that fixed costs are usually spread over period T while variable costs are allocated as units of output are produced.

Consider, for example, a process with an expected life of 10,000 units of output or 5 years whichever occurs first. Similarly assuming there are two plants, A and B, using the process. Given plant A produces 2,000 units per year, its end of life based on both output and time obsolescence will occur simultaneously at the end of 5 years. If plant B produces only at half the rate, say 1,000 units per year, then its process will have reached obsolescence after 5 years after producing only 5,000 units of cumulative output. In Figure 7.7, this simple problem is presented in real terms without discounting. With a cumulative output of 10,000 units over 5 years, the learning curve A_2B_2 includes fixed costs OC_2, while if the process is used only for a cumulative output of 5,000 units, fixed costs per unit output must be doubled to $OC_1 = 2xOC_2$, with a resulting increase in the total cost of production or learning curve to A_1B_1 as shown, where $A_1A_2=C_1C_2=OC_2$.

Assuming undiscounted variable costs of the Qth unit are

$$C(Q) = BQ^{-b}$$

where B is the variable cost of the first unit and Q are the cumulative numbers of units produced, then the discounted variable costs of 10,000 units produced over 5 years by plant A can be approximated by piecewise linearization as:

$$VC(A) = \sum_{t=1}^{5} \frac{B(2000t-1000)^{-b}\ 2000}{(1+i)^{t-0.5}}$$

and by plant B:

$$VC(B) = \sum_{t=1}^{5} \frac{B(1000t-500)^{-b}\ 1000}{(1+i)^{t-0.5}}$$

Similarly undiscounted fixed costs per unit are total fixed costs FCT divided by the number of units produced. In discounted terms, total fixed costs for plant I are:

$$FC(A) = \sum_{t=1}^{5} \frac{FCT/5}{(1+i)^{t-0.5}} = FC(B)$$

Or, as shown in Figure 7.7, total undiscounted or discounted fixed costs of the two plants are equal as both plants use the process for the same 5 years. Similarly if the process is used at a rate of 10,000 units per year and is worn out at the end of that year, the discounted cost of the cumulative production would be

$$\frac{FCT}{(1+i)^{0.5}} + \frac{B(10000-5000)^{-b}\ 10000}{(1+i)^{0.5}}$$

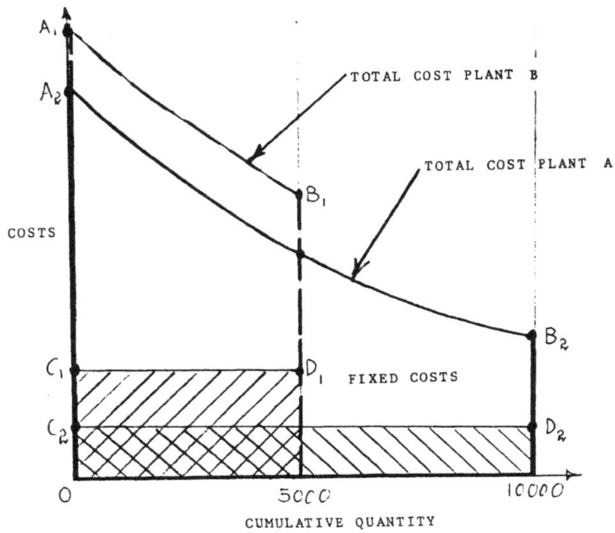

Figure 7.7. Effect of Rate of Output on Cost of Production.

The piecewise linearization can obviously be refined by monthly discounting.

In the simple analysis above, learning was assumed not to be affected by the rate of production or time and to be a direct function of cumulative output. Recent studies indicate that time as well as cumulative output affect variable costs or learning. These effects appear to be complex. With both too high a rate of production as well as an uncommonly low rate of production learning seems to decline. In the first case, probably because of the inadequacy of time for learning or feedback of experience, while if the production rate is very low, then the transfer of the experience is often lost because of lack of continuity.

So far we have assumed that all costs could be quantified and separated into fixed and variable costs. This is usually a fair assumption, if training, installation, run-in and starting costs can be included in fixed costs. Worker redeployment, retraining, casualty, and unscheduled repair costs on the other hand are difficult to capture as fixed costs and may have to be estimated and included under operating costs.

Qualitative costs, which may be caused by environmental impacts, changes in working conditions, performance risks, reorganization, and more, are usually not as easily incorporated into either fixed or variable costs as they are functions of output, time, as well as several other factors. To cope with qualitative costs (as well as benefits) the utility of specific items or qualitative benefits is usually estimated in a way which their allow adjustment as conditions change. For example, the quantitative fixed and/or variable cost of an alternative technology which is considered to be half as environmentally attractive as another, may be increased by say 20% to assure consideration of such important qualitative factors.

Although benefits of a process technology in general do not normally experience learning, they do exhibit both output and time dependency. For example the price of output is usually a function of cumulative output (with or without diffusion) and will normally decline as output increases, although other factors such as market share, age or obsolescence of technology, competing capacity and competitor's costs, as well as various qualitative factors play an increasingly important role.

Although it is difficult to separate benefits into fixed (time dependent) and variable (output dependent) benefits, such dual dependency is evident from empirical data.

Figure 7.8 shows the price curves for our two plants and the resulting net cumulative profit (benefit). The net present or discounted value of the net profit of these plants is

$$NPV(A) = 2000 \sum_{t=1}^{5} \left[\frac{A_1(2000t-1000)^{-a_1} - B(2000t-1000)^{-b} - FCT/2}{(1+i)^{t-0.5}}\right]$$

and

$$NPV(B) = 1000 \sum_{t=1}^{5} \left[\frac{A_2(1000t-500)^{-a_2} - B(1000t-500)^{-b} - FCT/5}{(1+i)^{t-0.5}}\right]$$

where for plant A price of the Qth unit produced is A_1Q and for plant B it is A_2Q.

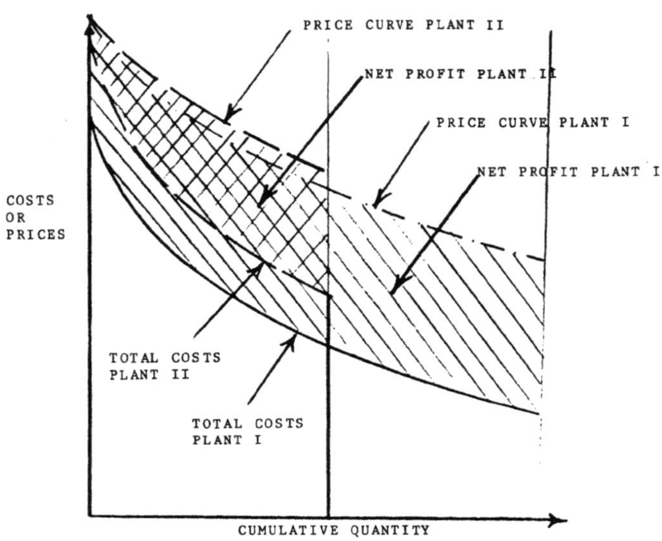

Figure 7.8. Net Cumulative Profit (Benefit) of Plants with Same Obsolescence But Different Cumulative Output.

Negative exponential price behavior is commonly observed in many products, though the slope of the price curve often changes as a result of increased competition, large-scale capacity addition, and/or introduction of newer technology.

7.4.1 *Objectives of Technological Change.* The objectives of technological change differ widely and may include quantitative measures such as:

1. net profit maximization in real or present value terms;
2. maximization of internal rates of return, return on investment, or return on sales;
3. minimization of investment payback period; and
4. maximization of market share or other measures of financial performance.

In many technological change decisions though economic performance, which often includes qualitative factors such as environmental impact, is of increasing importance.

Financial measures such as Net Present Value (NPV), Internal Rate of Return (IRR), or Discounted Benefit/Cost (B/C) ratio are often selected to determine comparative performance of alternative technological choices, timing of technological change, and rate and/or scale of introduction of or transfer to a new technology. For each decision set of alternative choice, timing, and rate and/or scale of introduction of a new technology, the benefit (revenue, etc.) and cost streams are computed over the expected cumulative output and/or lifetime horizon. Future expected benefits and costs are discounted using an appropriate (constant or variable) discount factor to obtain the NPV, IRR, or B/C for the particular decision set.

The set with the highest value of the chosen measure is then assumed to represent the most effective decision. It is important to remember that the decision set with the highest NPV may be different from that with the highest IRR or B/C, as shown in Figure 7.9 which shows that decision set I has a higher NPV than decision set II up to a discount rate of A, although decision set II has a higher IRR as shown, when NPV is zero or

$$0 = \sum_{t=1}^{n} \frac{(B_t - C_t)}{(1+IRR)^t}$$

where n = time horizon or process and B_t, C_t = benefit or cost during period t.

The IRR contrary to the NPV is a ratio measure and not a numeric. As a result, it ignores the scale of a process both in terms of rate and cumulative output. The same holds true for B/C which is usually expressed as:

$$B/C = \sum_{t=1}^{n} (B_t/(1+i)^t / \sum_{t=1}^{n} (C_t/(1+i)^t)$$

or the ratio of the NPV of Benefits to the NPV of Costs.

As a result, IRR or B/C tests of goodness or measures of performance should only be used in comparisons of technology choice decisions where scale, rate of introduction, time of introduction, and life of technology use (cumulative output and time) is the same for all alternatives considered.

7.4.2 *Technological Change Decision Problems.* Technological change decision problems often include not only choice, timing, scale, and rate of introduction of a new technology, but may also consider repetitive change of process technology, the zero technology change, process shutdown, and other options. Technology change decisions may also couple scale, capacity utilization, or rate of output to the price commanded by the output at the time it is produced, and/or the share of the market achieved (which in turn may affect the price). As a result, the technology change decision problem can become quite complex if modeled realistically.

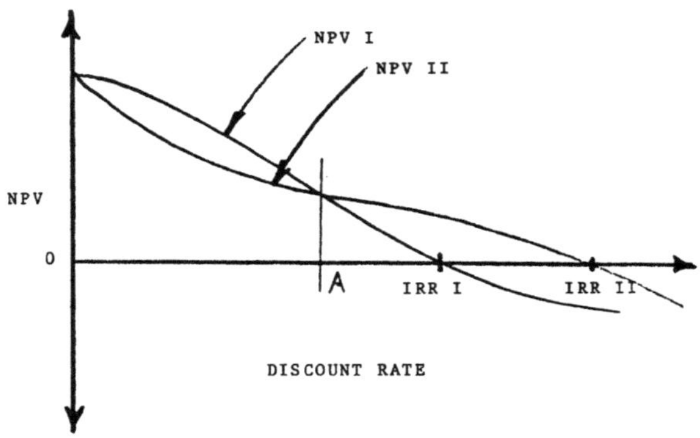

Figure 7.9. Comparison of NPV and IRR.

Considering a very simple problem as an example of producing 1000 units of output at the rate of 100/year by:

A. Use of the existing process which has a production cost of $C_1(Q+1000) = A_1(Q+1000) = 50(Q+1000)$ where $Q=1000$ have already been produced, but the process is capable of producing another 1000 units.

B. Introduction of a new process with a production cost of $C_2(Q) = A_2Q \quad = 60Q$

Assuming the price for output similarly has a negative exponential distribution

$$P(Q) = BQ^{-b} = 50\ Q^{-b}$$

and that a discount rate of 10% applies, the problem can be solved, assuming year end discounting (instead of mid-year discounting used earlier).

Average unit costs and price during a period of 1 year and annual output of 100 units is assumed to be that of the 50th unit produced. In other words

$$C_1(t) = \text{Average cost/unit of alternative A during period t} = 50\ (1000 + 100t - 50)$$

and

$$C_2(t) = \text{Average Cost/unit of alternative B during period t} = 60\ (100t - 50)$$

and

$$P(t) = \text{Average price/unit during period t} = BQ^{-b} = 50\ (100t - 50)^{-b}$$

The results are shown in Table 7.1. The differences in the NPV, IRR, and Benefit/Cost ratios should be noted.

Year	Average Unit Cost Alternatives A	B	Average Price	Net Profit/Unit Alternative A	B	Total of Discounted Net Profit Alternatives A	B
1	.271	1.919	2.990	2.719	1.071	247.201	97.396
2	.253	.730	1.356	1.103	.626	91.122	51.734
3	.238	.466	.939	.701	.473	59,645	35.537
4	.225	.346	.737	.512	.391	34.978	26.664
5	.213	.277	.615	.402	.338	24.955	20.934
6	.202	.233	.532	.330	.299	18.605	16.899
7	.193	.201	.472	.279	.271	14.295	13.901
8	.185	.177	.426	.241	.249	11.230	11.592
9	.177	.159	.389	.212	.230	8.974	9.766
10	.170	.144	.359	.189	.215	7.269	8.294
Net Present Value						511.274	292.717
Net Present Value Cost Alternative A							136.026
Net Present Value Cost Alternative B							354.583
Net Present Value Revenues							647.300
Benefit/Cost Ratio Alternative A = 4.75							
Benefit/Cost Ratio Alternative B = 1.82							

Table 7.1. Simple Cost/Benefit Example.

The effect of a changeover from an existing technology, say plant A, to a new technology, say plant B, can also be shown on a log-log plot (Figure 7.10) which indicates the effect of the price curve on profit when the technology is changed after plant I has produced 100 units. In that case the price for the 101st unit produced by plant I or the first unit produced by plant II would be OA or OB. As a result, plant II would incur a loss (area ADF) initially followed by profit DEC^2, versus a profit of BGHC^1 if technology of plant I is retained. A more complicated problem may involve not only choice of alternative technologies but also different timing and rate of production as pointed out before.

In other words, we may have to compute the outcome of the technological change decision under all possible choice, timing, and rate of production alternatives.

Considering choice and timing as our decision variables and assuming for example the choice of the existing technology I and two new alternative technologies, II and III, as well as introduction of one of the new technologies now when I has produced Q_1 units, and II as well as III have produced no units ($Q_2 = 0 = Q_3$), or wait until unit I has produced $Q_1 + q$ units where $q = 100$, 200, or 300, or 400, and 100 units are required per year for the next 5 years. This problem can be solved recursively by dynamic programming as follows.

We define the five decision periods (beginning of each year) as stages 1 to 5 and the technological choices as states I, II, and III, as shown in Figure 7.11.

The assumption is made that once a change is made, the chosen technology will be used to produce all remaining required output. This means that the technology can be changed from I to II or III at any stage, but not from II or III once a change has been made. This assumption could obviously be relaxed to allow any stage to be performed by any technology, obviously assuming proper consideration of production costs at that state based on previous experience of the particular technology.

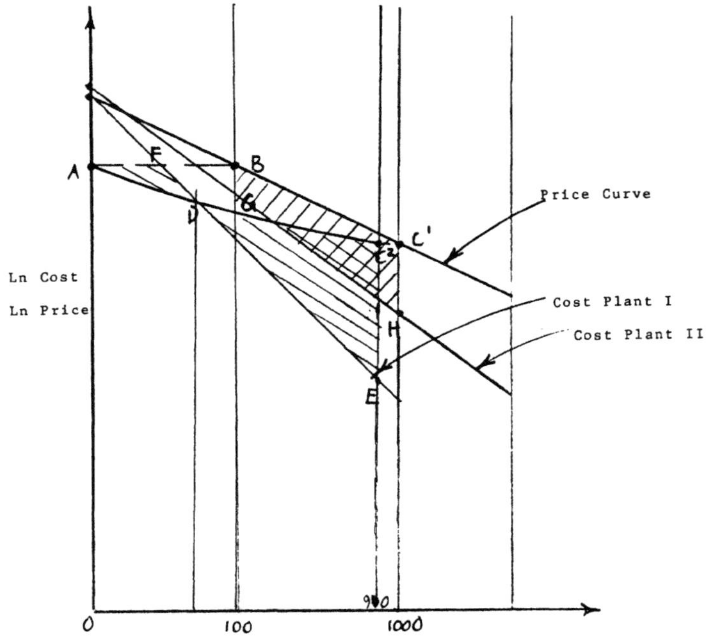

Figure 7.10. Effect of Technology Switch (Without Discounting).

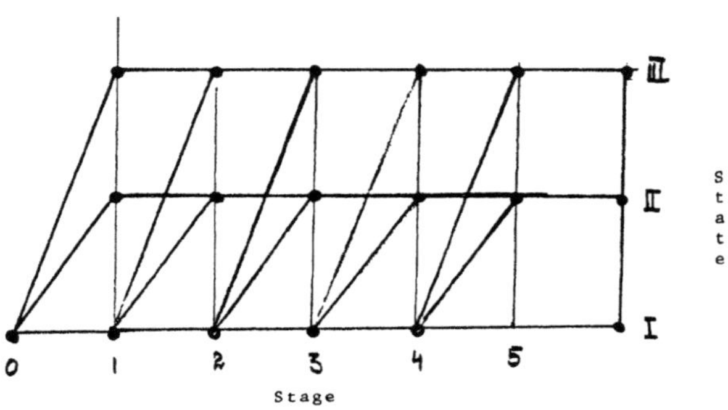

Figure 7.11. Decision Alternatives - Three Technologies and Five Decision Points.

If S_n is the state (process) used in stage n, then a decision x_n moves the process to some state s_{n+1} in stage n+1. At each decision point (beginning of a period of year), a decision can be made.

Let us define $f_n(i)$ as the minimum cost of the n stages, given we used process i in stage (n+1) from the end. The NPV or discounted cost of using technology i during stage n is $DC_n(i)$ and for the whole 5-stage decision process:

$$
\begin{aligned}
f_5(i) &= \text{Min } [DC_5(i) + f_4(i)] \\
f_4(i) &= \text{Min } [DC_5(i) + f_3(i)] \\
f_3(i) &= \text{Min } [DC_3(i) + f_2(i)] \\
f_2(i) &= \text{Min } [DC_2(j) + f_1(j)] \\
f_1(i) &= DC_1(i)
\end{aligned}
$$

If we want to restrict the technology change to only one change decision, then the above general model is constrained by allowing a change only once or if for $f_4(i)$, $f_5 = i = 2$ or 3, then $S_4 = S_3 = S_1 = i$ or the state remains constant at j for all subsequent stages.

Similarly if for $f_3(i)$ $S_4 = i = 2$ or 3. $S_3 = S_2 = S_1 = i$ or the state remains constant for all subsequent stages. The general model can obviously be used if the $DC_n(i)$ is adjusted for the experience attained by technology i if assigned to stage n. Dynamic programming can be used for problems with many more decision variables (stages and states).

When different technologies and rates of production are used, various feasible combinations of technological choice and rate of production constitute particular states at each stage over time (or adjusted cumulative production).

References

1. Spence, A. M., "The Learning Curve and Competition", Bell Journal of Economics, Vol. 12, 1981.

2. Lieberman, M. B., "The Learning Curve and Pricing in the Chemical Industries", Rand Journal of Economics, Vol. 15, No. 2, 1984.

3. Spence, A. H., "The Learning Curve and Competition", Bell Journal of Economics, Vol. 12, No. 1, 1981.

4. Sultan, R. G. M., "Pricing in the Electrical Oligopoly", Vol. 2, Division of REsearch, Graduate School of Business Administration, Harvard University, 1975.

5. Wright, T. P., "Factors Affecting the Cost of Airplanes", Journal of Aeronautical Sciences, Vol. 3, No. 4, 1936.

6. Stobauch, R. B. and Townsend, P. L., "Price Forecasting and Strategic Planning - The Case of Petrochemicals", Journal of Marketing Research, Vol. 12, 1975.

7. Shwartz, L., Horesh, R., and Roz, B. "Trade in Technology: Management Decisions and Pricing", Technological Forecasting and Social Change:, Vol. 23, 1983.

8. Technological Forecasting

Throughout human history, and particularly in the last few centuries, technology has been the dominant force creating changes in people's lives. Yet, it is only recently that managers in private and public organizations have realized the need for forecasting or projecting technological change and the resulting impact on their activities as well as the environment in general. To be useful, technological forecasts need not necessarily predict the precise form technology will take or a given application for some technology at some specific future date. Like any other forecasts, their purpose is simply to help evaluate the probability and significance of various possible future developments so that managers can make better decisions.

Since technology is not simply knowledge of physical relationships, but includes ways of performing different activities, technological forecasting can range from projection of the initial glimmerings of how a basic phenomenon, inventions, or discoveries can be applied to the solution of a problem, but similarly include projections of innovative approaches to development of technology, improvements in the design of a product, device, or production process, as well as forecasts of new types of applications of existing technology.

The performance of any process, product, or operating system is normally improved in small, often continuous increments over time as innovation proceeds. What may appear to be a "step function" advance in a technology is usually nothing more than an accumulation of small advances not worth introducing individually until they additively make a significant change in the total technology. Moreover, a given technology generally includes a variety of competing designs, each with a distinctive balance of performance and economic characteristics which appeals only to certain people. Finally, of course, a specific process or product in a technology may also fulfill quite divergent needs and perform very dissimilar functions for its various owners.

It is the basic continuity in a technology's technical and economic characteristics and its potential applications which makes technological forecasting desirable. Except over short terms, when immediate direct extrapolations of present techniques may be useful, it is futile for the forecaster of technology to predict the precise nature and form of the technology which will dominate a specific future application. But, he can make "range forecasts" of the performance characteristics that a given use is likely to demand in the future. He can make probability statements about what performance characteristics a particular class of technology will be able to provide by a certain future date. Also, he can analyze the potential implications of having these technical-economic capacities available by the projected dates.

A variety of techniques for technological forecasting have been developed. As in all other forecasting methodologies, those most effective, are based on careful analysis of past experience combined with the study of the factors influencing future developments in technological change and the insights of competent and imaginative people experienced in the technology of interest and knowledgeable in the uses or markets of the technology. Technological forecasting requires observation and measurement of underlying data, trends, and in particular, analysis of technology developments.

There are many promising technology forecasting techniques available, which are general approaches or techniques rather than actual formal or methodological methods.

Technological forecasting is intimately related to technological planning. Technological planning is usually interpreted as the development of an effective program which uses technological forecasts for future decision. In recent years, the development cycle or time between generations of new technology has shortened appreciably. In the past when the time between technological generations was equal to or larger than the normal economic or operating life of technology, technological decisions could be based exclusively on present technology and past experience. Today the technological cycle time has shortened to but a fraction of the normal lifetime of most technologies. As a result, it is now imperative to consider the projected state of development of future technology to guide today's technology decision. Such forecasts of future technology should not only consider the next generation of the technology of interest, but also technological forecasts of related technological developments, particularly those with potential technological interfaces which means technology with similar characteristics and functions though different applications or technology which falls between the technology of interest and other technologies.

Economic forecasts are usually based on well founded theories. In fact, satisfactory forecasts based on historic data are among the main tests of economic theory. Technological forecasting, on the other hand, relies less on extrapolations from the past to project future developments than on current status, work in progress, and future development environments. It assumes that any technological development, even in a completely unrelated field, may have an effect on the future of technology of interest. In recent years development of technological forecasts have increasingly included consideration of environmental, social, political, military, legal, and economic factors. Such technological forecasts are often performed by statistical evaluations of responses from an experienced group of experts selected by statistical methods to reduce bias. These responses are generally in the form of answers to detailed and well developed questionnaires. One such technique which is finding increasing applications in technology forecasting is the 'Delphi' technology forecasting method or technique.

The Delphi technique elicits opinions from a number of experts with the aim of generating effective group response. This is usually performed by a carefully planned, anonymous, structured program of sequential and repetitive interrogation by questionnaire. In this manner convergence of opinion is encouraged by feedback of anonymous group opinion. This and similar approaches are of particular value in technology forecasting where a general knowledge of current status, trends in research and developments, as well as a perception of existing voids in science and technology must be used to predict the future. As technological developments occur more frequently and rapidly now, and as the impact of such technological developments affect an ever larger segment of the population to an increasing degree, technological forecasting is now increasingly being elevated from an academic exercise to an essential economic and management planning tool. In addition to the Delphi technique, various other trend, impact, and other forecasting methods have been developed for technology projections. The selection of technique depends on various factors, such as the history of the technology of concern, if any.

Trends in technology are influenced by many factors as shown in Figure 8.1. Among these are:

1. scientific and technical discoveries and inventions;
2. advances in innovation;
3. limit analysis;
4. discovery of technological voids;
5. technological incentives;
6. demand forecasts and assessments;
7. competitive threats; and,

8. opportunity analysis.

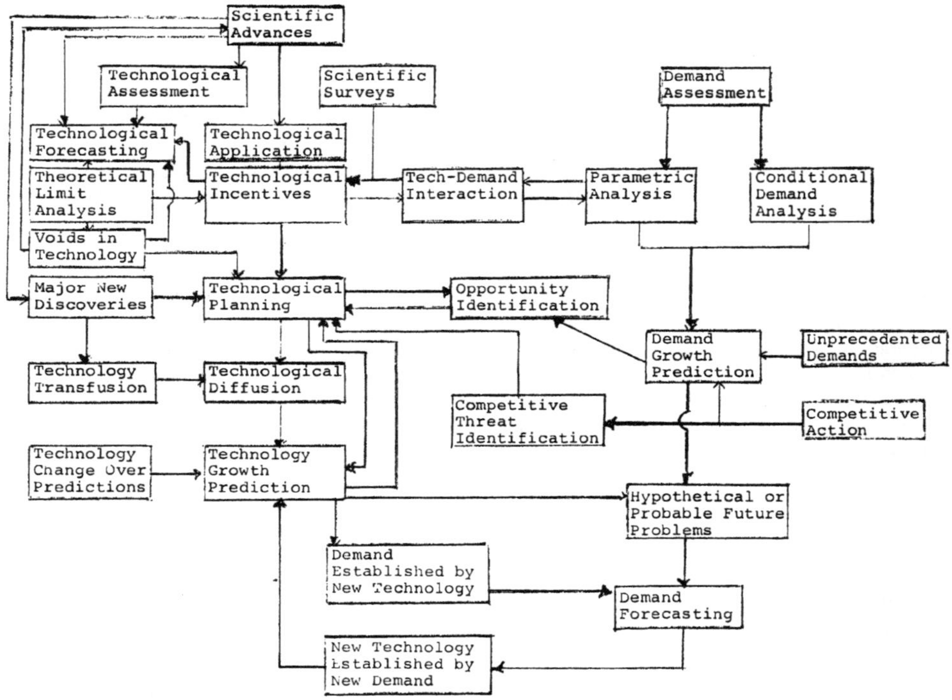

Figure 8.1. Trends in Technology.

All of these factors may play a role in the development of new technology or new applications of technology, and should therefore be considered or included in the technological forecast. Technological planning is usually concerned with a more limited horizon, and performed for a firm, defined group, or society. It uses as inputs more specific demand analysis and forecast opportunity identification, projected payoff determination, competitive threat identification, projected payoff determination, competitive threat identification, resource limitation, know-how surveys, and evaluations of technological transfusion and diffusion capability. Overriding considerations are usually financial, economic, social, political, and strategic need (or goals) ordered by some priority system.

It is often desirable to improve our technological forecasts by a better understanding of the process of technological advancement. This can be done by establishing various "cause and effect" relationships which in turn permits a better understanding of the mechanisms of technological advance and transfer, and of the forces which drive such developments.

The technological forecasting assessment and planning process also requires an evaluation of the effect of system parameter changes, functional and operational changes and the availability of subsystems or components which make the technological advance real or useful.

Both the benefits and costs of alternative technological approaches as well as resulting resource allocation can be estimated by identifying the functions the technology would be expected to perform in the future. As opportunities for technological progress occur more frequently, it is increasingly important to integrate technological forecasting assessment and planning. As noted before, forecasting methodology must include a detailed understanding of the process of technological advancement and technology transfer. Similarly careful validation should be performed of both the objective and the information inputs. It is usually advantageous to incorporate probabilistic or conditional probability density estimates into the technology forecasting, and use conditional, stochastic forecasting models.

The objectives furthermore should, if possible, be expressed in terms of economic, social, and political utility. Another important consideration is the effect of feedback. Technological developments usually open up new opportunities which in turn induce additional resource allocation for other technological advances and/or technology transfer. A typical approach to technological forecasting and planning is shown in Figure 8.2.

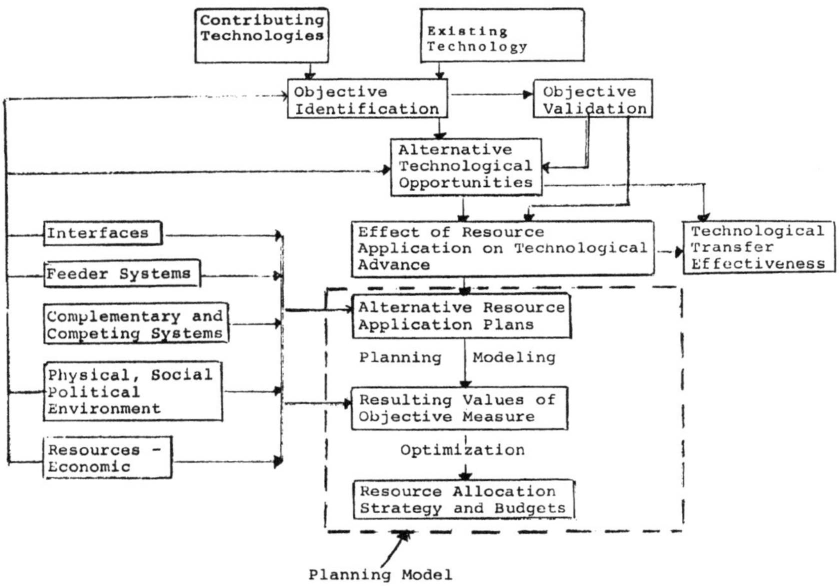

Figure 8.2. Technology Forecasting and Planning Model.

The inputs into such a model may, in addition to system particular requirements, consist of information on such factors as:

1. availability (non-availability) of technology;
2. cost (economics) of use of current technology;
3. interface effectiveness;
4. intermodal integration;
5. changes in capital intensity, investment distribution,

and use of resources, manpower (skill);
6. technology transfusion and diffusion;
7. capability of technology absorption;
8. public image, interest, and incentive;
9. safety, reliability, cost factors;
10. environmental factors; and,
11. competitive factors.

It is important to chose a forecasting approach which permits consideration of all the factors which are expected to influence future developments of the technology of interest.

8.1 METHODS AND TECHNIQUES OF TECHNOLOGICAL FORECASTING

Various analytical techniques are available for technological forecasting. These include:

1. mapping methods
2. network analysis such as relevance tree diagrams, decision or conditional networks, decision trees, graphic network models;
3. systems analysis models;
4. demand assessment models;
5. limit analysis methods;
6. changeover point prediction techniques;
7. matrix methods;
8. polling of experts and statistical opinion analysis; and,
9. statistical data analysis methods.

Effective hierarchical structuring and use of subjective matrices which translate objectives of research, development, and innovation, and their relative values into outcomes of alternative technological developments are often useful approaches to force explicit statements about such objectives. The inputs to a technology planning model are usually in the form of quantifiable performance parameters, which are used to derive the resulting values of resource requirements, resource schedule, and level of objective measures. The most important decision in technological forecasting and planning concerns the determination of the uncertainties involved in the cost, time, development, transfer, application, acceptance, and operational success of new technology. It is for this reason that technological forecasting and planning cannot be performed effectively as a once-through process. It requires continuous feedback and updating which in turn provide the inputs for improved estimates of future uncertainties in technology developments. With all the shortcomings inherent in any attempt of planning for the future, based on forecasts derived from insufficient data and knowledge, it is increasingly important to use formal technological planning approaches not only to eliminate inaccurate planning but to reduce the probability of downright mistakes.

Technological change decisionmaking is less and less a 'seat of the pants' art. It now requires a scientific process which includes estimates of the probabilities of various subjectively determined outcomes. In order to estimate the outcome of technological developments, it is necessary to develop a seasoned view of the future context in which new technological developments will be implemented. These anticipated states of affair constitute projections. The projection which appears most probable is then called a technology forecast and adopted as the basis for decisionmaking.

Besides providing the basis for rational technological decisionmaking, forecasts enable the manager to assess the impact of future developments, their incidences, and magnitude. Forecasts also allow him to evaluate the less tangible aspects of the impacts of technological change. This is often done by allowing comparison between projected technological developments in the absence of any program with post program states which provides at least an estimate of program 'benefit'.

Projections and forecasts may also serve a heuristic function by highlighting the limits to the technological choice of alternatives. By illuminating the probable/possible causes and effects they assist the management of technological change decisionmaking process.

Technological forecasting techniques fall into three groups: (1) time series and trend projections, (2) model building and (3) simulation and qualitative forecasting. There are many methods available, some of which span more than one of these groups.

The first group includes trend extrapolation, regression, and probabilistic forecasting, and is largely based on historical data that is analyzed in various statistical ways to arrive at forecasts of the future based on the past. These techniques are the most commonly used, but may not be very useful in technological forecasting where historic developments often play a small role in future developments.

The second group includes dynamic models, quantitative, cross-impact analysis, and a computer simulation version called KSIM, input-output analysis, and policy capture. These methods are based on models or simulations of the technological change phenomena to be forecasted. They demonstrate the interactions of the separate factors. Such models are helpful in attaining a broad perspective and a better grasp of the totality of the problems, in foreseeing effects that might otherwise be overlooked in evaluating the prospects of future technological developments.

The last group includes Delphi, expert opinion and cross-impact methods, alternative futures, and values forecasting. This kind of forecasting is more global and subjective. It can be both qualitative or quantitative.

A combination of several of the above methods is often found to be useful. Technological forecasting should not only be based on past and current trends, but should also reflect:

- significance, and effect on systems, of the interaction among selected prime parameters affecting technological performance and the development of new technology - interaction with other technical areas and intensity of effort in these areas;
- effect of private and public investment;
- effect of political and military interests;
- effect of national and international laws such as those pertaining to safety, pollution, waste disposal, and more;
- effect of labor, social management, and administrative factors.

Technological change is intimately linked to societal developments which, in turn, are greatly affected by forces of change. In some societies these forces are an omnipresent force, while in others they arise occasionally, only to become dormant again. Much depends on the goals of a society as fostered by its culture and nurtured or curtailed by its political structure and leadership. Societies develop value systems which play a role in the ranking of priorities and ultimately value systems.

In designing a technological forecast we must remember that technological developments do not take place in a vacuum but in an economic, physical, social, and political environment which is continuously changing, in part because of the impact of technological developments themselves. We therefore deal with a feedback system in which technological changes are fostered, directed, or induced by an economic, physical, social, and political environment which, in turn, is always affected by the technological developments which impact on the culture, economy, social life, and political structure of the same society. This cyclical behavior is now pervasive and the feedback requirements have become shorter and shorter, and permeate most aspects of a society's activities, structure, institutions, and moral values as well as goals.

Technological forecasting must therefore consider the environment in which the technology is to be forecasted. During time of war and emergencies for example societies change their priority and value systems which will influence the selection of technologies designated to receive more attention and resources for more rapid advancement and change. Radar, jet propulsion, and similar technologies which are important to defense were all developed during war time. In several cases, the initial technological discovery was actually ignored until its defense value was recognized, when it suddenly received great support. Effective technological forecasting should therefore be designed to include a feedback or cross-impact cycle which allows introduction of the impact of changing conditions on technological change and vice versa for the duration of the planning horizon. In other words, the impact of possible exogenous developments, such as war, economic alliances, etc., on the rate and type of technological development should be included in the technology forecast.

8.2 THE DELPHI AND OTHER SUBJECTIVE OR EXPERT OPINION TECHNIQUES FOR TECHNOLOGICAL FORECASTING

Subjective methods of forecasting technological change and development have become not only popular but the most important means for technological projection and planning. [1, 2, 3, 4]

8.2.1 *The Delphi Technological Forecasting Technique*. The "Delphi" method is a systematically converging opinion survey technique for the development of a consensus of experts on future technological developments. It is based on sequential opinion surveys where the survey questions are increasingly more specific, with specificity affected by developing consensus. The method can be shown to provide better forecasts than those obtained from individuals or even pools of individuals.

A Delphi forecasting 'experiment' is usually performed by the use of carefully designed questionnaires issued to a select yet broad group of 'experts'. After each round of questionnaire responses, questionnaires are reworked using some of the information obtained from the answers, including opinion, feedback, and supported by relevant information, either volunteered by respondents or obtained otherwise. This process of sequential questionnaire opinion polling is continued until a reasonable degree of consensus is obtained.

The results of a Delphi experiment may not only identify technological trends and development alternatives, but also provide insights into the assumptions and methods used in deriving judgements by the selected experts. The technique is also educational because the built-in feedback provides the respondents with information and opinions which may be useful in forming their informed judgements. A typical Delphi technological forecasting experiment consists of the following steps:

1. Definition of process, product, or service technology to be forecast. This often takes the form of definition of a technological problem or opportunity and not of a particular type of product or process with some general performance characteristics.

2. Determination of the areas, range, and depth of expertise, and knowledge required to fairly judge future technological developments.

3. Identification and evaluation of experts in terms of credentials, credibility, reputation, communications skills, and bias.

4. Selection of respondent group (anonymous).

5. Identification of major issues affecting technological developments under consideration, including constraints, exogenous, regulatory, and institutional issues.

6. Preparation of a first questionnaire. This should be broad or general and elicit respondent comments, information disclosure, and more.

7. Analysis of questionnaire results and development of general consensus including elimination of some considerations and developments. Formulation of additional feedback and other information designed to force convergence of expert opinions. Tabulation of results.

8. Preparation of second questionnaire and distribution.

9. Repetition of step (7) and test for consensus or at least adequate convergence of opinion for a reasonable forecast of technological change.

10. Questionnaires are reissued until such consensus is reached, when final responses, as well as background information, are compiled, and the results formulated as a technological forecast.

A Delphi experiment may result in more than one alternative technological forecast. It is important to keep the objective of the forecasting experiment well in mind , but at the same time assure that experts are not influenced by the objective or objectives. To guard against this, potential Delphi experiments usually use 2 or more separate groups of experts, all made up of members of users, monitors, and experts. In other words, experts in a group are supplemented by representatives of potential users or beneficiaries of the future technological developments.

Monitors, on the other hand, are individuals who are actually involved in the design of the experiment. One problem faced by Delphi experiments is how to assure anonymity of the respondents, how to assure timely and full responses, the valuation of convergence, and therefore determination of the number of questionnaire cycles required, and the decision of whether to retain identical respondent groups throughout the different questionnaire cycles.

The Delphi approach seeks to obtain a group opinion through an anonymous, multi-level group interaction. The mechanisms which produce an answer in the classical Delphi are normally called a "conditional scientific prognosis": if certain preconditions hold and if certain laws and rules are valid, then certain predicted events will occur with such and such probability.

The panelists are (in the ideal case) experts in the special field of application and their number has to be big enough to cover all aspects of the topic and will therefore depend on the complexity of the questions and on the numbers of related fields. Anonymity is necessary to guarantee that the ideas and arguments are not influenced by the reputation of the panelists supporting them and that there will be no cooperation and coordination of the panelists during the Delphi inquiry. Usually a Delphi exercise involves separate groups of individuals and different roles for these groups, such as:

1. User group which consists of individual(s) who expect some sort of advantage from the exercise.

2. Design and monitor group which designs the initial questionnaire, summarizes the returns, and redesigns the follow-up questionnaires.

3. Respondent group chosen to respond to the questionnaires. This may sometimes be the user body or the user body may be a subset of the respondent group.

Other issues are: (1) the anonymity of the respondent group among its own members to the design or user group; (2) the time allowed the respondent group to answer; (3) the number of necessary iterations; and, (4) the use of different respondent groups in each iteration. In general, a Delphi technology forecast has no hard and fast rules to guide its design, and the success of the Delphi is dependent upon the ingenuity of the design team and the background of the respondent group. The utility of the results depends upon the close cooperation between the design team and the intended user body or at least a clear understanding by the design team of the goals or requirements of the user body. The Delphi requires a degree of quantification to be imposed upon subjective judgemental factors and the definition of this quantification is a matter of principal concern to the design team.

There[1] are a lot of variations on the Delphi technology forecasting technique ("classical Delphi") such as more specialized Delphi applications for policy forecasting and decision prediction, as described in Table 8.1.

The "Policy Delphi" method was developed as a conditional scientific prognoses are very rare in the social sciences and more "long-term" prognoses were required. An important factor to recognize here is that technology and policy Delphi are methods for predicting future developments and not to influence reality or conditions.

Several extensions to the classical Delphi technique have been introduced to increase the technique's utility and/or efficiency. These extensions include: cross impact analysis; SEER (system for event evaluation and review); and, integration of trend extrapolation with Delphi.

[1] This section is adapted from "Management of Technological Change in the Shipbuilding Industry", E. G. Frankel, A. H. Aslidis, and Ping Ho, MIT Ocean Engineering Report, 1985.

	Classical Delphi	Policy Delphi	Decision Delphi
Background			
The significance of reality	Reality is given; its interpretation is clear; consequences are discussed.	Reality is given; its interpretation will be discussed	Reality will be created.
Goal			
Delphi serves as a forum for	Facts	Ideas	Decisions
The procedure tries to	Create consensus	Define and differentiate views	Prepare and support decisions
The basic concept is	Conditional scientific prognosis	Pluralism	Self-fulfilling prophecy
Panelists			
The panelists are	Unbiased experts	Lobbyists	Decision makers
The panelists try to	Obtain realistic statements and prognoses	Support and succeed with their standpoints	Create a basis for realistic and useful decisions
The participation has to	Be high in absolute terms (i.e., many panelists)	Consider all relevant groupings	Cover a high percentage of the relevant decision makers
Method			
The feedback serves for	Obtaining the realistic answer of prognosis	Getting well-defined group opinion	Stimulation and information of the decision makers
Anonymity means that	The participants in the inquiry are not known and all answers are anonymous		The participants are known from the very beginning, the answers are nevertheless anonymous
The reason for the anonymity is to	Hinder arrangements and personal influences	Facilitate extreme viewpoints and objectivity	Support personal answeres and to raise the participants
The strict objectivity of the evaluation has	Mainly methodological reasons (to be unbiased)	Mainly pragmatical reasons (to get a complete picture)	Mainly ethical remains (the director of the study must and influenced the decision process)

Table 8.1. Classical, Decision, and Policy Delphi: A Comparison.

Source: Stanford "Forecasting Techniques", Stanford University, March 1982.

Cross-impact analysis was developed to solve the problem of interrelationships among forecasted events. In other words, forecasts which consist of collections of potential future occurrences, likely dates of their occurrence, and the probability of their occurrences might contain neutral reinforcing or mutually exclusive events. The exclusion of cross effects is the key variable in cross-impact analysis. Usually cross effects are determined by consensus. Consistency of forecasts results is achieved through a series of reviews in which the evaluations of the cross-impact matrix is revised until consistency is obtained. Eventually, a complete matrix is developed which indicates modes of linkage between events, strengths of relationships, and predecessor-successor orientations, or relationships.

The SEER technique was developed to make Delphi forecasting more efficient by generating initial lists of events through interviews, and by clearly indicating a set of rules, which reduces the amount of time required by the Delphi panelists. SEER limits the required time by constraining the forecast to two rounds and requesting that panelists answer only those questions within their areas of expertise. SEER also addresses the question of event desirability from a user's point of view and event feasibility from a producer's point of view. Finally, utilization of this technique produces a Delphi forecast in which interrelationships, goals, supporting events, and alternative paths are clearly specified.

The third extension of the Delphi technique is an integration of cross-impact analysis, SEER and trend extrapolation. The new factor in this approach is the application of Delphi to the determination of the future trends. Panels are given the historic (time series) data and asked questions designed to elicit their opinions concerning the future environment and direction of trends. The design and results of this process are similar to SEER and cross-impact matrices.

8.2.1.1 *Propensity for Opinion Change in Delphi Exercises.* Some psychological processes are involved in the formulation of a result from a Delphi procedure. The two most important processes involved are:

1. the psychological factors involved in opinion change, and,

2. those factors apparently related to the individual's decision.

The feedback information provided to the participants at each round is essentially a measure of central (concentrating) tendency. These measures act as external psychological anchors; that is, they are stimuli exerting a relatively large influence on the determination of judgement. In continuing a Delphi exercise, an individual may respond in one of three ways:

- ignore the anchor
- shift judgement away from the anchor, perhaps in the interest of moving the anchor closer to their true desire; (contrast) or
- shift judgement towards the anchor ("assimilation").

The process of assimilation has been much studied in the literature. Many psychological experiments concerning attitude change involve participants who are being provided with an anchor. Any resulting attitude change is examined in light of the effectiveness (credibility of the anchor). The presence of an anchor implies the potential presence of "cognitive dissonance" which in turn may lead to assimilation. In general, agreeing with people decreases dissonance while disagreeing increases it.

Since traditionally Delphi studies have employed experts or participants, one would presume that the credibility of the participants (panelists) would be high. This might not be the case for a non-expert group. The result would be perfect correspondence in any subsequent round. On the other hand, a perfectly incredible communicator would exercise no impact on subsequent rounds. The credibility of most panels would be somewhere between the two extremes.

For mildly credible anchors, at low and moderate discrepancies, the curve possesses a positive slope; however, at high discrepancies the curve turns downwards. This can be explained because low discrepancies are likely to fall within an individual's latitude of acceptance, inducing assimilation, while this is not the case at high discrepancies where "contrast" is more likely to occur. For highly credible communicators, the latitude of acceptance for an individual will be increased so that some moderate movement will occur (although evidence suggests that at high discrepancies there would be a condition of neutrality).

To terminate a Delphi forecasting experiment, the procedure is usually to set aside questions for which the evaluation of the respondents reflects agreement or consensus. Questions for which no agreement is reached are explored further in subsequent rounds of interrogation.

Consensus is assumed to have been achieved when a certain percentage of responses fall within a prescribed range for the value being estimated. Problems arising from the use of such stopping criteria for Delphi studies stem from the fact that measures that determine the dispersions of a group response are neither necessarily, nor strictly, measures of consensus. Depending on the limits used in setting the stopping criterion, "consensus" may or may not describe the actual level of agreement reached by the respondents.

Even when the criteria is an adequate indicator of consensus, there are no guidelines concerning the strategy to be followed when stable responses are obtained, nor are there guidelines to be followed when no consensus or when low levels of agreement are obtained.

To develop a stopping criterion that explicitly differentiates the concepts of stability and agreement in Delphi studies, it is necessary to define these concepts and the relationship between them.

Stability refers to the consistency of responses between successive rounds of a study. It occurs when responses obtained in two successive rounds are shown statistically not to be significantly different from each other, irrespective of whether a convergence of opinion occurs. On the other hand, different levels of agreement (convergence of opinion) among respondents may occur in any given round, irrespective of whether the round is stable when compared to the preceding one. While stability does not necessarily imply a given level of agreement, it is only when a stable answer is reached that the analysis at the level of agreement should be attempted. Hierarchically, the most important concept is that of stability, since an unstable answer that might be obtained in a given round has no value in the final analysis. The sequential analysis of the extent of stability and agreement reached in a given round will then lead to the determination of the strategy to be followed in the succeeding rounds of the study. So long as an unstable response is obtained for a given question, additional rounds of the Delphi study must be performed. The levels of agreement or outcomes can be are defined as follows:

a. consensus - unanimity concerning the issues of concern is achieved when the experiment is terminated;

b. large majority - more than 70% of the respondents exhibit consistency on the issues of concern and the experiment is terminated;

c. A simple majority of respondents exhibit consistency. Experiment may be terminated or continued.

d. equality - respondents are equally divided over the issue. In this case the experiment group structure may be changed or issue rephrased, and experiment continued until majority agreement is obtained.

e. plurality - a larger portion of the respondents (but less than 50% reaches agreement, in which case those disagreeing can be regrouped and experiment continued until majority agreement is obtained;

f. disagreement - each respondent maintains views independent of the other respondents, such that the responses cannot be brought into consensus. Here an effort at rephrasing may be attempted, but if this is found unsuccessful, the experiment would be terminated anyway.

The χ^2 - Test of Stability

A stopping criterion that measures whether a stability of response has been achieved, and that seems to be consistent with the rules of the statistical theory, is one that utilizes a non-parametric Chi Square test to check for the independence of the rounds from responses obtained in theory. The conduct of such a test does not require the assumption of a particular probability distribution of responses.

In order to conduct a Chi Square test, the population and sample are classified in terms of two attributes. For a Delphi study these attributes are the specification of the round and the type of response. Responses must be classified into two or more intervals, and the number of responses in each interval for each round must be seven. The χ^2 test then proceeds to test the null hypothesis (H_o) against an alternative hypothesis (H_1) defined as follows:

H_o = the Delphi rounds are independent of the responses obtained in them
H_1 = the Delphi rounds are not independent of the responses obtained in them.

If H_o turns out to be true, stability has been relieved while more rounds are necessary if H_1 is shown to be true.

The mechanics of conducting the test of these two hypotheses involves the computation of χ^2-statistic and the comparison of its value with a critical value obtained from standard statistical tables. To calculate this statistic both the observed and expected frequencies of occurrence of particular types of responses in each of the two rounds being tested must first be determined. The former are readily available from the raw Delphi data, and the latter are determined under the assumption that the null hypothesis is true. For a contingency table displaying the frequency responses for each response interval in each of two consecutive rounds, the expected frequency that would occur in each cell if the null hypothesis was true would be the average of the two frequencies observed in the two consecutive rounds. So:

$$\text{Chi Square} = \chi^2 = \sum_{i=1}^{2} \sum_{j=1}^{N} \frac{(O_{ij}-E_{ij})^2}{E_{ij}}$$

where O_{ij} and E_{ij} the observed and expected frequencies by which the respondents select a response intense j in round i.

The critical value of can be selected from standard tables for a given level of significance and so we can reject or accept hypothesis H_o.

Indeed, it has been demonstrated that individual stability in consecutive Delphi rounds implies group stability, but not conversely; and that a test for individual stability on Delphi studies will provide more information than a test for group stability.

8.3 CROSS-IMPACT FORECASTING OF TECHNOLOGICAL CHANGE

Future technological developments invariably involve uncertainty and incorporate implicit and sometimes explicit assumptions about the occurrence and interactions of future events on each other and thereby on the factors which form the basis for future technological developments. Furthermore, prediction of technological change, which always involves a multiplicity of variables, should be realistically represented by complex models which show all the events which potentially affect developments and consider their interdependencies in causing or reacting to other events. Because the actual process by which technology develops is complex, we usually make grossly simplifying assumptions and truncate variables considered in the technological projection, even if we are fully aware that they have a major impact on the projection of interest.

Another major problem is that in forecasting technology by quantitative trend or similar analysis based on past developments we concentrate on one or more variables which are readily quantified (and which we assume to be independent), while excluding variables which may be more important but which are subjective in nature, difficult to define and quantify or highly complex.

The cross-impact method provides an approach for the investigation of the interdependencies between such variables and between the occurrence and non-occurrence of an event at a specified period of time as well as at subsequent time periods, as a result of different developments.

In essence, cross-impact technology forecast analysis allows a prediction of future technological developments by a method which includes allowances or considerations of all the interdependent factors which influence the future development of the technology forecasted, including those variables which cannot be readily quantified.

Since it was first proposed by Gordon and Hayward [5] the cross-impact method has been widely studied and applied to long-range planning as well as demand and technology forecasting. In the cross-impact analysis, the following sequence of steps are used:

1. Definition of the Technology to be Forecasted

Events which can be expected to affect the development of the technology of interest are first identified, including the conditions and causes which may influence their occurrence.

2. Identification of Variables

Next we identify the variables which define the above events. These variables can be qualitative or quantitative and can be either subjective or objective in nature. Each of the variables is next given upper and lower limits.

3. Set-up of Cross-Impact Matrix

A cross-impact matrix is established which shows the interactions between the different events represented by their associated variables as shown in Figure 8.3. The entries in the matrix can be positive, negative, and zero, with positive entries indicating that the causal event enhances the affected event, while a negative entry expresses an inhibiting impact.

The cross-impact matrix can be deterministic with entries indicating a definitive causal impact as expressed by the cross-impact coefficient. The cross-impact coefficients can also have associated conditional or transition probabilities. Similarly occurrence of causal events may be uncertain, when each such event has an associated probability of occurrence. Finally, cross-impact matrices like input-output tables are often time variant, and are developed for each of the future time periods of interest separately, as the cross-impacts associated transition probabilities, and probability of occurrence of causal event may all change with time.

The estimates of the time varying cross-impacts (as well as their associated conditional probabilities and probabilities of causal event occurrence if the forecast is to be based on probabilistic and not deterministic inputs) are usually derived by a Delphi-type consensus seeking approach.

4. Evaluation of Alternatives

When a consensus is reached, the evaluation of alternatives commences. The forecasters inspect the impacts produced and consider the consequences of alternative policies and test the sensitivity. Specially, a critical development or event may be tested for impact or for sensitivity, by being replaced by another event or by changing the initial values of the event. Thus used, the cross-impact method becomes a way to examine the interrelationships between objectives and strategies.

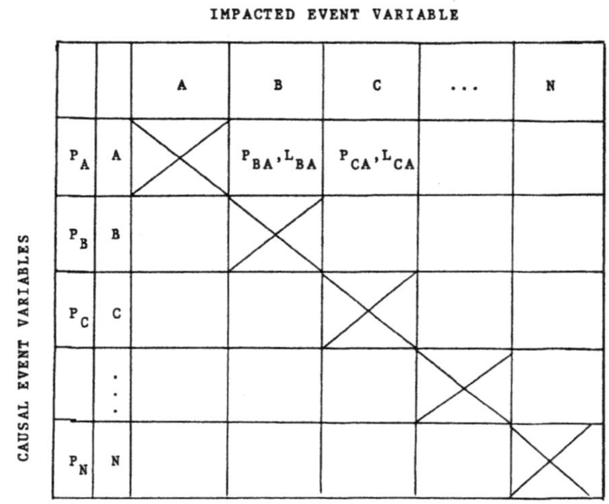

Figure 8.3. The Construction of the Cross-Impact Matrix.

Considering cross-impact analysis further, let us assume that a number of technological forecasts or developments which are expected to potentially affect future technological developments are available in terms of their probability of occurrence, probability distribution of the time of their occurrence and probability distribution of their size. If we designate these causal developments or events as D_1, D_2 ... D_n with associated probabilities P_1, P_2 ... P_n and if initially the variables representing these developments are taken as deterministic and if these developments are assumed to have an impact on each other and on the technologies to be forecasted, then a cross-impact matrix could be formed which relates these impacts and defines the interactions. Interactions are of various types. A development D_K may be unaffected by the occurrence of or non-occurrence of another development D_m. Similarly occurrence of a development D_K may eliminate the occurrence of another development D_m or it may be linked and require the occurrence of development D_m.

If P_{mn}^1 = probability of D_n occurring after D_m has occurred

then $P_{mn}^1 = f(L_{mn}, P_n, t_m, t_n)$

where

L_{mn} = interaction between developments
P_n = probability that D_n will occur before occurrence of D_m is considered
t_m = time in the future of D_m

t_n = time occurrence of D_n

then assuming a quadratic relationship

$$P_{mn}^1 = a\,P_n^2 + b\,P_n + c = a\,P_n^2 + (1\text{-}a)\,P_n$$

where $0 < a < 1$ if D_m decreases the probability of D_n occurring and $-1 < a < 0$ if D_m increases the probability of D_n occurring. If developments D_n and D_m do not affect each other's probability of occurrence but do affect the technological change by the respective developments, similar inhibiting and enhancing functions can be established.

The coefficient "a" can be assumed to be expressed approximately by:

$$a = KL_{mn}\,\frac{t_n - t_m}{t_n}$$

where

K is +1 or -1

depending on the mode and L_{mn} is a number between 0 and 1 depending on the strength of the interaction. (Perfect interaction for example will result in $L_{mn} = 1$ while zero interaction in $L_{mn} = 0$).

As a result we can now express the conditional probability

$$P_{nm}^{1} = KL_{mn}\left(\frac{t_n - t_m}{t_n}\right) P_n^{2} + \left[1 - KL_{mn}\left(\frac{t_n - t_m}{t_n}\right) P_n\right]$$

The above relations establish the conditional cross-impact probabilities which can then be multiplied by the unconditional technological change projected to be generated by the respective developments to obtain total technological change generated by these developments.

Developments or events expected to affect the technological change of interest and projected to occur within the forecast horizon must therefore be identified first and their probability of occurrence estimated. In the simplest case their size and degree of interaction by each development, etc. are assumed to be deterministic. The conditional probabilities are next computed and the resulting technological development probabilities are determined.

8.3.1 *KSIM-QSIM.* Simulation languages designed to facilitate the application of cross-impact analysis are available under the names KSIM and QSIM. Unlike basic cross-impact analysis the simulation models stress the structural dynamics of the interactions among developments and the relations among their interactions. This permits quantitative and qualitative interactive factors to be introduced more effectively than the approach taken by the basic cross-impact method which relies primarily on the determination of future development probabilities. The principal characteristic of KSIM is that it permits the use of any type of relevant data from subjective estimates of interactions to quantitative linkages between developments.

Technological change is usually affected by many developments, too many for normal human comprehension and use in the application of traditional forecasting techniques. As a result, most technology forecasts are based on a highly truncated set of considerations, often simply historic trends, modified slightly to incorporate some already known development. Simulations generally require numerical inputs and deterministic or probabilistic coefficients in the simulation model. On the other hand, subjective factors such as acceptance of technological change are seldom included in analysis using simulation. Similarly simulation models are usually highly structured and assume rigid strategies which reduce the ability of decisionmakers to test alternative scenarios with east.

KSIM [6] was designed to permit quantitative and qualitative factors to be considered in a simulation model which represents continuously changing interactions among all developments which are expected to affect future developments of technology of interest.

In KSIM variables are bounded and change with the impact of other variables. The change of a variable in response to impact by other variables tends to zero as the variable approaches its upper or lower bound (sigmoidal character). In general, variables will produce larger impacts on technological developments (the system of interest) as they grow larger. Most interactions among variables are described by looped networks (in matrix or network form).

Using the approach of Kane [8] and assuming $x_i(t)$ to be a bounded variable designating say normative traffic generated by development i at time C

$$0 < x_i(t) < 1;\ 1,2,\ldots N \text{ and } t \geq 0$$

and

$$x_i(t+ t) = x_t(t)P_i(t)$$

where

$$P_i(t) = \frac{1 + \frac{\Delta t}{2}\sum_{j=1}^{N} (|x_{ij}|) - A_{ij})x_j}{1 + \frac{\Delta t}{2} \Sigma\ (|x_{ij}| + A_{ij})\ x_j}$$

and

A_{ij} = interaction or impact of x_j on x_i
= period of one time iteration

In simpler form we can express

$$P_i(t) = \frac{1 + \Delta t \mid \text{sum of negative impacts on } x_i}{1 + \Delta t \mid \text{sum of positive impacts on } x_i}$$

As $t \rightarrow 0$ the above expressions can be expressed as differential equations

$$\frac{dx_i(t)}{dt} = - \sum_{j=1}^{N} A_{ij}\ x_i(t)\ x_j(t)\ \ln x_i(t)$$

The impact coefficients A_{ij} need not be constants. They may be functions of time or state variables.

Technological forecasting may be concerned with projecting technological developments in a global, industrial, regional, local, or individual firm (even a single person firm) environment. It may similarly consider a particular narrowly-defined process or product technology or a general process or product technology. The time horizon of the forecast may also be a fixed future time, a period to a particular time in future, or indeterminate time.

Considering a simple problem of projecting future change in process technology for a firm over a fixed period of time or a one step (period) forecast of technology adoption we first identify the causal and affected events as well as their expected developments. Assume there are two possible process technologies which could be introduced, and that these affect investment costs, productivity and process applicability.

If we use an integer scale from -2 to +2 to show the cross-impacts with -2 implying most inhibiting, -1 inhibiting, 0 no effect, +1 enhancing effect, and 2 major enhancing effect and if our expert opinion consensus has obtained the following impacts of improvements in productivity from 20 manhours/unit all the way to 12 manhours/unit, with investment costs from a low of $1 million to a high of $5 million.

Applicability	Productivity (In Manhours/ Unit Output)*	Investment Cost (in $ Millions)**	Range of Impacts
0 - 25%	18 - 20	4 - 5	(-2)-(-1)
25% - 50%	16 - 18	3 - 4	(-1)-(0)
50% - 75%	14 - 16	2 - 3	(0)-(+1)
75% - 100%	12 - 14	1 - 2	(+1)-(+2)

* Highest possible productivity 12 manhours/unit.

** Lowest investment cost $1.0 million.

and the following cross-impacts were agreed upon: for two alternative technologies I and II. Applicability here means the range of types of outputs that a technology can produce.

	Tech I	Tech II	Applic.	Prod.	Cost
Tech I	1	-1	0	1	-2
Tech II	+1	1	-0.5	2	-1
Applic.	0	0.5	1	0.5	-1
Prod.	-1	-2	-0.5	1	+1
Cost	+2	+1	+1	-1	1

This simply means that technologies I and II are assumed to enhance themselves and slightly inhibit each other's developments (cross-impacts of +1 and -1 respectively). Similarly technologies I and II are expected to have a 50% and 37.5% applicability, a productivity of 14 and 12 manhours per unit output, and an investment cost of $5 million and $4 million respectively. We can next determine which of the two technologies would be developed under normal circumstances using the methods presented in the Appendix to this chapter.

8.3.2 *Recent Improvements and Revisions of the Cross-Impact Method.* Duperrin and Godet [7] developed an approach which aims at assuring the consistency if the estimated probabilities of given events, in which a consistent probability system is constructed by use of a quadratic programming technique, which takes higher order probabilities into account as well. Assuming n events e_i (i=1,2,...n), and

P(i,t) = probability of occurrence of e_i during period t
P(i,t/j) = probability of occurrence of e_i during period t given event e_j occurred
P(i,t/j) = probability of occurrence of e_i during period t given event e_j did not occur

also

$$\hat{0} \leq P(i) = (i,t) \leq 1$$

and for consistency

$$P(i,t/j) = P(i/j) \text{ and } P(i/j)P(j) = P(j/i)P(i) = P(i,j)$$

$$\text{and } P(i/j)P(j) + P(i/j)P(j) = P(i)$$

If x_i is a decision variable which is 1 if e_i occurs and 0 if it does not occur, and $E_K = (x_1, x_2 ... x_n)$ = state of event vector = scenario K where K = 1,2...2^n. Similarly $P(E_K) = P(x_1, x_2 ... x_n)$ = probability of scenario K. To assure consistency of the joint probability P(i,j) and P(i,j) the authors [6] suggest the use of a quadratic equation

$$J = \sum_i \sum_j [P(i/j)P(j) = P^*(i,j)]_2 + \sum_i \sum_j [(Pi/\bar{j}) = P^*(i/\bar{j})]_2$$

which should tend towards min $P(E_K)$ and is subject to

$$P^*(i,j) = \text{Consistent } P(i,j) = P[x_1, ... x_i=1, ... x_j=1, ...]$$

$$P(i,j) = P(x_1, x_2 ..., x_i=1, ... x_j=0, ...]$$

$$\sum_1^n P(x_1, ... x_n) = 1$$

and $P(x_1, ... x_n)$ 0.

Another simpler approach is to minimize

$$J = \sum_i \sum_j [\underline{P}(i/j)\underline{P}(j) - \sum_{K=1/2n}^{2^n} t_{ijk} P(E_K)]_2$$

$$+ \sum_{i,j}[P(i/j)P(j) - \sum_{K=1}^{2n} S_{ijk}P(E_K)]_2$$

again subject to

$$\sum_{K=1}^{2^n} P[E_K] = 1 \text{ and } 0 \leq P(E_k) \leq 1$$

when

$$P^*(i) = \sum_K X_{iK}P(E_K)$$

where

x_{ik} = 1 if e_i occurs as part of E_K
x_{ik} = 0 if e_i does not occur as part of E_K

Similarly

$$P^*(i/j) = \sum_{K=1}^{2^n} t_{ijk} \underline{P}(E_K)/\underline{P}^*(j)$$

where

t_{ijk} = 1 if e_i and e_j occur as part of E_K
t_{ijk} = 0 otherwise

and

$$\underline{P}^*(i/j) = \sum_{K=1}^{2^n} S_{ijk} P(E_K)/1-P^*(j)$$

where

S_{ijk} = 1 if e_j occurs and not e_i or vice versa as part of E_K
S_{ijk} = 0 otherwise

Mitchell [8.9] and later Kaya et al [10] showed that the quadratic programming model does not provide unique n dimensional scenario probabilities, as J becomes semidefinite as n becomes larger. They also point out that intuitively the first and second order probabilities could realistically not determine higher order probabilities. In addition, it is argued that the quadratic programming formulation imposes uncommonly large computational requirements.

One issue which the above formulation does not consider explicitly is the precedence of events e_i and e_j which is reviewed by Mitchell et al [11]. Kaya et al [10] propose a revised cross-impact method in which the maximum and minimum value of each scenario probability, $P(E_K)$ is evaluated by solving a linear programming problem.

f = Min or Max $P[x_1...x_n]$

subject to $P^*(i) = P(E_K) = P(x_1,...,x_i=1,...x_n)$

Similarly the computational effort is reduced by decreasing the number of events simultaneously considered and by making the problem a multi stage problem.

Finally, instead of estimating the conditional probabilities (i/j), estimates of impact probabilities are derived, which are defined as the probabilities that e_i occurs during the time interval, given e_j occurs at the beginning of the time interval (P(i j). The procedure is divided into stages. The time intervals between the stages, the set of events considered in each (or various) stages are first defined.

First Stage

1. Next exogenous conditions which occur during the first stage but are expected to affect future events are identified.

2. Event e^1_i out of the i=1...n events identified which have a high probability of occurrence in stage 1 are selected.

3. The event probabilities P(i) and the impact probabilities P(i/j)=P(i j) are then estimated for the first stage.

4. The factors

$$D_i = \Sigma|\underline{P}(i\ j) - \underline{P}(j)|$$

and

$$A_i = \Sigma|\underline{P}(j\ i) - \underline{P}(i)|$$

are next calculated to obtain an estimate of the total impact of e^1_i to and from other events respectively.

5. A limited number of events whose impacts, D_i and A_i, are relatively large are next selected.

6. Calculate two-dimensional probability, P(i,j) for these events.

7. Based on the estimates of the probabilities P(i) and P(i,j), construct an n-dimensional consistent probability set by use of a quadratic programming, where n_i is the number of events.

$$J = \Sigma_i U_i[((i) - P^*(i)]^2 + \Sigma_i\Sigma_j\ V_{ij}[P(i,j) - P^*(i,j)]_2 \rightarrow \min,$$

where U_i and V_u are appropriate weighting parameters.

8. By applying linear programming, determine a dimension of scenarios and upper and lower bounds of the scenario probabilities, corresponding to P*(i) and P*(i,j) given at the preceding step.

9. Select m_i-dimensional scenarios, (S^1_j, j = 1,...,K_1] whose minimum probabilities of occurrence are large. They are considered highly likely scenarios.

Next Stage

Execute the following procedure for each likely scenario S^1_j.

10. Identify the events for the next stage, which comprise those events not included in n_1 events of the first stage and those events that do not occur during the first stage. Determine exogenous conditions for the next stage.

11. Repeat stage one and let the resulting scenarios be

$$[S^2_{ji}, i= 1,...,k_j].$$

12. Construct a set of total scenarios

$$S_K = [S^1_j \vee S^2_{ji}]$$

where v denotes a combination of scenarios of different stages.

Cross impact matrix technology forecasting has become an increasingly valuable tool and has been applied to the forecast of transport, power generation, electric power transmission, and other technological developments.

Technological discoveries or innovations in progress are usually identified which might have an effect on the technological developments to be forecasted. It is often also useful at that stage to identify new modes or applications of existing technology and possible changes in user preference and societal constraints which may affect technological developments. In other words, it is preferable to forecast technological developments addressed to the solution of a problem than the emergence of a particular technology such as high temperature superconductive materials.

Dates of occurrence and initial probabilities are estimated for each event in the stages of development of the various technologies identified (simulating the results of Delphi panels for example). This then leads to the construction of a matrix which shows the predecessor-successor relationships as well as the cross-impact mades of linkage strength of developments on each other as well as the various exogenous and endogenous factors on the developments of interest. The matrix is run until reasonable convergence is achieved. It is often interesting to study the inferences which can be drawn from such a cross-impact technology forecasting exercise. One is the test of sensitivity of shifts in probability of future technological developments to changes in the initial probability levels.

An interesting and instructive example of this cross-impact forecasting technique is presented in Ref. [10] in which the interrelationships between connected events is presented by a formal analytical technique.

References

1. Fusfeld, A. R. and Foster, R. N., "The Delphi Technique: Survey and Comment - Essentials for Corporate Use", Business Horizons, June 1971.

2. Linstone, H. A. and Turaff, M., "The Delphi Method: Techniques and Applications", Addison-

Wesley Publishers, Reading, MA, 1975.

3. Raugh, W., "The Decision Delphi", Technological Forecasting and Social Change", Vol. 15, 1979.

4. Riggs, W. E., "The Delphi Technique: An Experimental Evaluation", Technological Forecasting and Social Change, Vol. 23, 1983.

5. Gordon, S. and Haywood, H., "Initial Experiments with the Cross Impact Matrix Method of Forecasting", Futures, Vol. 11, 1968.

6. Kane, J., "A Primer for a Cross-Impact Language - KSIM', Technological Forecasting and Social Change, Vol. 4, 1972.

7. Duperrin, J. C. and Godel, M., "SMIC - A Method for Constructing and Ranking Scenarios", Futures 7, Vol. 4, 1975.

8. Mitchell, R. B., "Scenario Generation Limitations and Developments in Cross-Impact Analysis", Futures 9, 1977.

9. Mitchell, R. B. and Tydeman, J., "Subjective Conditional Probability Modeling", Technological Forecasting and Social Change, Vol. 1, 1978.

10. Kaya, Y., Ishikawa, M., and Mori, S., "A Revised Cross-Impact Method and Its Applications to the Forecast of Urban Transportation Technology", Technological Forecasting and Social Change, Vol. 14, 1979.

11. Novaky, E. and Korant, K., "A Method for the Analysis of Interrelationships between Mutually Connected Events: A Cross-Impact Method", Technological Forecasting and Social Change, Vol. 1, 1978.

Technological Forecasting Bibliography

1. Alexander, A. et al, "Measuring Technological Change", Technological Forecasting and Social Change, Vol. 5, pp. 189-203, 1973.

2. Bardecki, M. J., "Participants' Response to the Delphi Method: An Attitudinal Perspective", Technological Forecasting and Social Change, 25, 281-292, 1984.

3. Blackman, A., "A Cross Impact Method Applicable to Forecasts for Long-Range Planning", Technological Forecasting and Social Change, Vol. 5, pp. 233-242, 1973.

4. Blackman, A. W., Jr., "The Market Dynamics of Technological Substitutions", Technological Forecasting and Social Change, 6, 41-63, 1974.

5. Fusfeld, A. R. and Foster, R. N., "The Delphi Technique: Survey and Comment - Essentials for Corporate Use", Business Horizons, June 1971.

6. Kwasnicka, H., Galar, R., and Kwasnicki, W., "Technological Substitution Forecasting with a Model Based on Biological Analogy", Technological Forecasting and Social Change, 23, pp. 41-58, 1983.

7. Linstone, H. A. and Turoff, M., The Delphi Method: Techniques and Applications, Addison-Wesley, Reading, MA, 1975.

8. Raugh, W., "The Decision Delphi", Technological Forecasting and Social Change", 15, pp. 159-169, 1979.

9. Riggs, W. E., "The Delphi Technique: An Experimental Evaluation", Technological Forecasting and Social Change, 23, pp. 89-94, 1983.

10. Sharif, M. N. and Sundararajan, V., "A Quantitative Model for the Evaluation of Technological Alternatives", Technological Forecasting and Social Change, 24, pp. 15-29, 1983.

11. Turoff, M., "A Synopsis of Innovation by the Delphi Method", Talk presented at the 1970 ORSA Meeting in Detroit, Michigan.

12. Turoff, M., "The Design of a Policy Delphi", Technological Forecasting and Social Change, 2, pp. 149-171, 1970.

13. Gordon, S. and Haywood, H., "Initial Experiments with the Cross-Impact Matrix Method of Forecasting", Futures, Vol. 1, 1968.

14. Kane, J., "A Primer for a Cross-Impact Language - KSIM", Technological Forecasting and Social Change, Vol. 4, 1972.

15. Kane, J. et al, "A Revised Cross-Impact Method and Its Application to the Forecast of Urban Transportation Technology", Forecasting and Social Change, Vol. 14, 1979.

APPENDIX 8A - EXAMPLE OF A CROSS-IMPACT TECHNOLOGY FORECAST - SHIPBUILDING

Manufacturing in shipbuilding is undergoing rapid change as more low-cost labor countries compete with traditional shipbuilders in industrial countries. As a result, major technological changes in formerly labor-intensive activities such as metal cutting and welding are under development and consideration, including completely integrated automation and robotization of cutting and many of the fitting and welding processes.

Various scenarios of shipyards of the future are envisioned and analyzed using the revised cross-impact forecasting technique. The technology forecast is divided into two stages, short-term technological change from present to 1992 and long-term technological change from 1992 to 2000.

Various events under consideration during these two stages are listed in Table 8A-1 and their level of application (1-5) is denoted, where 1 implies scarcely used and untried, while 5 implies general acceptance by advanced shipbuilders.

Number	Technological Event	Level of Stage 1	Application Stage 2
1	Computer-aided design (CAD) and automated control of individual cutting and welding processes	3	5
2	Computer-aided design and manufacturing (CAD/CAM) of individual cutting and welding processes	2	4
3	Integrated automated cutting	3	4
4	Integrated automated cutting and welding	1	4
5	Laser-controlled fitting of cut plate into two-dimensional sections	3	4
6	Laser-controlled fitting of cut plate into three-dimensional blocks	3	3
7	Integration of fitting controls with welding controls		2
8	Completed integrated automated cutting, edging, fitting, and welding and block manufacturing assembly		2

Table 8A-1. Technological Events in Shipbuilding

Step 1

The effect of technological changes considered on labor (manhour) requirements in steel fabrication during the first stage which are considered to affect the future of shipbuilding in industrial countries are estimated by a group of experts.

Step 2

Events 1-6 in Table 8A-1 are selected for the first stage.

Step 3

P(i) and P(i j) are estimated for the chosen events with results shown in Table 8A-2, which also displays the values of D_i and A_i computed as explained in the text.

i \ j	1	2	3	4	5	6	D_i	Ranking
1	.45	.35	.75	.89	.30	.50	.06	6
2	.35	.40	.74	.91	.29	.50	.13	5
3	.50	.25	.75	.85	.15	.30	.60	2
4	.15	.33	.65	.90	.25	.30	.72	1
5	.45	.40	.75	.90	.30	.65	.15	4
6	.46	.40	.78	.88	.50	.50	.26	3

Table 8A-2.

Step 4

Events with largest impacts are next chosen (we choose all the six events).

Step 5

Determination of two-dimensional probabilities [P(i,j)] obtained from P(i) and P(i j). (Table 8A-3) A method to compute P(i,j) from P(i) and P(i j).

i \ j	1	2	3	4	5	6
1	.450	.151	.348	.373	.125	.227
2	.151	.400	.270	.356	.118	.200
3	.348	.270	.749	.651	.197	.350
4	.373	.356	.651	.899	.265	.416
5	.135	.118	.197	.265	.300	.211
6	.227	.200	.350	.416	.211	.500

Table 8A-3. Probabilities P(i) and P(i,i) - First Stage.

Step 6

Determination of consistent probabilities [P*(i)] and [P(i,j)] by use of quadratic programming (Table 8A-4).

i \ j	1	2	3	4	5	6
1	.449	.151	.348	.373	.134	.227
2	.151	.399	.270	.356	.118	.200
3	.348	.270	.799	.651	.197	.350
4	.373	.356	.651	.898	.265	.416
5	.134	.118	.197	.265	.300	.211
6	.227	.200	.350	.416	.211	.499

Table 8A-4. Consistent P*(i), P*(i,j) - First Stage.

Step 7

Evaluation of minimum probabilities to reduce event set.

Step 8

Selection of most probable scenario (5) and determination of scenario dimension (number of events) (Table 8A-5), that is exclusion from further consideration of the events that affect very little the others.

Events Occurring	Events Not Occurring
3	1
4	2
	5
	6

Table 8A-5. A Likely Scenario.

This completes the first stage. In the second stage, all second stage events are considered and the procedure is repeated for each scenario of stage 1 that is considered probable. Now stage 2 contains all events not occurring in stage 1 (plus other possible events). In the above example when we consider the scenario of Table 8A-5, the second stage would include events 1, 2, 5, 6, 7, 8. Events 3 and 4 supposedly occurred during stage 1. This procedure is repeated for all stages. The number of scenarios, though, increases multiplicatively because for each scenario of stage 1 we associated a number of scenarios of stages 2 and so on.

9. Product, Process, and Service Innovation Cycles

Innovation is the most important part of the technological change process. While invention provides the spark for technological change, it is development of technology by innovation which provides the energy and allows the new technology to make a contribution. It is innovation which brings an invention into commercial use by converting it into a product, process, or service which can be used or applied and for which there is real demand. While ideas in terms of discoveries, inventions, or research results are necessary to start technological innovation, only ideas worthy of development will usually be subjected to innovation. In terms of investment, the large majority of expenditure in the development of new technology is spent for the innovation of the technology and not for research or discovery leading to the idea or invention. This ratio is usually 1 to 10 or higher.

The innovation process may start as soon as a technological discovery or idea is made, or it may start only after a long delay. The time lag between the discovery of the technology and the start of its innovation process depends on:

a. accessibility of idea or discovery,
b. ownership of idea or discovery,
c. potential market or demand for technology,
d. investment requirements for innovation of technology,
e. objectives of owner of idea or discovery,
f. availability and performance of competing technology,
g. identity of owners, producers, or users of competing technology, and
h. market share and cost of competing technology.

Once the innovation process is started, financed, and undertaken by the inventor and/or other owners (or some third parties), the rate of innovation are affected by factors such as:

1. potential profitability based on market acceptance, cost of competing technology and demand;
2. investment required and impact of investment into innovation on rate of development of technology;
3. producibility and resulting cost and price;
4. technical requirements;
5. organizational requirements;
6. management needs;
7. size of firm undertaking innovation;
8. innovation process scale effects;
9. standardization, regulation, and other constraints imposed;
10. rate of technological obsolescence of new as well as competing technologies;
11. user inputs; and,
12. input requirements.

Innovation is sometimes induced by exogenous developments such as new types of demands, shortage of competing technologies, and the need for substitute technologies, government regulations which reduce competition or use of other technologies, restrictions on the availability or use of inputs required for or by the competing technologies.

The effect of size of the innovator on the rate of innovation is more pronounced in some fields such as energy, pharmaceuticals, etc. On the other hand, larger firms are often not larger or more successful innovators in fields such as electronics.

In recent years, more inventions or ideas are acquired from external sources, particularly in the U.S., but there is a need for improvements in the methods of discovery and evaluation of external inventions or ideas to assure that acquisition of rights and investment into innovation can be made under reasonable terms that have a high probability of success. Evaluations required usually include

a. potential investments and other commitments required,
b. extent of uncertainty in the technological developments, usefulness, and marketability of new technology,
c. methods and degree of protection,
d. potential for early diffusion and/or introduction of new partners or users during innovation to reduce or eliminate capital exposure induced by acquisition and innovation/development costs, and
e. market and competing technology developments.

In theory, large firms try to be technology leaders and once committed introduce innovations faster. Yet this is not always the case as firms with older or conservative management and firms which suffer under liquidity shortages may drag their feet or diffuse technology before effective development.

Product and process technology develops in a cyclical manner with highly concentrated product and process developments, during one part of a cycle, usually followed by times of low development or consolidation. This phenomena is particularly pronounced during product or process innovation when development appears to be much more affected by exogenous factors such as market and economic conditions than during invention or discovery stages of technology. Similarly the rate of diffusion during different steps in the innovation stage depends largely on exogenous factors and is, as a result, quite cyclical.

A number of theories have been advanced to explain the behavior of technological developments during the innovation activity. Although endogenous factors such as technological capability play some role, the most important influence on the innovation process appears to be exerted by social, economic, political, and competitive developments. Innovation, even more than invention, is not the result of random genius activity performed sporadically or inexplicably. It is nearly always the result of environmental and other exogenous developments, amplified by endogenous factors such as marketing or financial objectives, role perception, and status preservation. Social evolution, need, or popular demand may spurn innovation as does often political pressure or strategic requirements. Most innovation, while triggered by microfactors, is sustained by macro consideration of a firm, region, nation, or world entity.

In other words, while initial stages of innovation are often just the carryover from the invention or discovery process, innovation soon languishes, unless other driving forces are generated, mostly exogenously. Often an incubation period is required to make exogenous factors aware of an innovation opportunity and to allow the technical preparations to be made for effective innovation. This may require lengthy negotiations, transfer of rights (by licensing or otherwise), and formal fiscal and technical planning, which often takes longer in the U.S. where legal issues are of greater concern and usually involve lengthy procedures, as compared with those required in countries in the Far East, for example, where such legal issues, as rights to an innovation, are often settled verbally. The incubation period in innovation is also often called a period of hesitancy during which the inventor hesitantly relinquishes development control of the discovery to innovators and their supporters.

The role of inventors in the innovation process is often difficult to assess or determine. On one hand their involvement is helpful in smoothing innovation, particularly during the initial stages of innovation, yet they may also interfere with orderly innovation if their objective or focus is distinctly different from that of the innovators or sponsors of the innovation process.

The skills and approach required for successful innovation are in many cases quite different from those used in discovery and exhibited by inventors. Few inventors are able to judge potential applications of an invention realistically and are willing to consider a broad based approach to innovation because discovery itself is usually a very focused activity. As a result, inventors may find themselves in conflict with innovators whose principal goal is to bring the invention into use of markets and for the purpose of the innovation sponsor.

The most important change during the product and process innovation cycle is the decreasing role of inventors in innovation, independent of the origin or personality of the inventor, who may be an employee of an industrial firm, a researcher in an R&D organization or university, a government official, or just a private citizen. Few inventors have the means to develop their discoveries and, if sponsored, are only seldom entrusted with the managerial responsibility for the innovation or invention development process.

Innovation is usually not restricted to a particular product or process, as product innovations may require process innovations and vice versa. Furthermore, both may demand service innovations to be effectively applied. As shown in Figure 9.1, product, process, and service innovation are interdependent and, as a result, overlap quite often. A new product may require different design, manufacturing, and testing, as well as a different approach to marketing and servicing. Conversely, a new process may allow new products to be introduced which in turn require a different type of service, all of which may result in interdependent produce, process, and service innovation.

The rate of product innovation, often measured in terms of expenditure per unit time, usually starts high and then declines gradually over the whole product innovation cycle. Conversely the related process technology innovation, designed to develop effective processes to manufacture or produce the new product usually starts at a low level and then peaks when most of the characteristics of the new product technology have been developed.

The innovation process is influenced by many factors. Successful innovation requires knowledge of technology and the market, and the ability to judge development potentials, market acceptance, and competition. Innovation is affected by the position of the potential product, process, or service generated by the invention in terms of use, market, and competition.

It is also a function of the sponsors' strategic objectives which may be to delay or prevent the emergence of a new product, process, or service because it would compete with an existing one in which the sponsor has a large investment.

Innovation is also affected by the cultural and structural environment in which it takes place. Some approaches used or innovation goals may, for example, be culturally unacceptable. The management of innovation is complex because it requires conciliation of often conflicting objectives. It is a dynamic activity, subject to ever changing internal and external conditions.

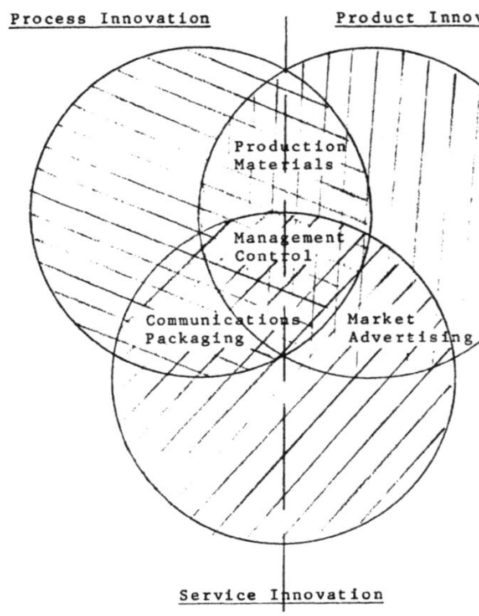

Figure 9.1 Product, Process, Service Innovation Interdependence.

Innovation also attracts other innovation and may cause a chain reaction, through parallel or competing product, process, or service developments. As an innovation in a field receives support and is developed, others will usually join in the development or invest heavily in the development of competing innovations, which may be improvements on existing products, processes, or services. This results in a rapid acceleration of the innovation process and even more investment in innovation for the particular purposes of the invention. Once a breakthrough is achieved and the innovation is applied and starts to be diffused, investment declines and drops to a level just above that before that innovation cycle started. As a result, investments in innovation are observed to be cyclical in most technological fields. Figure 9.2, for example, shows the investment cycles in shipyard automation innovation from 1960 to 1987. As noted, the frequency as well as the amplitude of the investment cycles changed significantly over this period, with a rapid decline in frequency and a large increase in amplitude of the innovation investment cycles. Similar behavior has been observed in other technologies. It appears that not only has the frequency of innovation increased, but the rate of innovation and the investment made in innovation has also grown significantly, both in real and current terms.

9.1 DIFFUSION OF INNOVATIONS

Diffusion of innovations can occur at any stage of the innovation process. It may take place when the invention is still a breakthrough in basic knowledge without any development or even known application or take place when the invention is fully developed for one or all its possible uses or applications. At each stage of the innovation process there are usually existing and potential adopters. The difference between the number of existing and potential adopters is a function of the availability of information on the innovation and the range as well as interest in possible applications of the innovation.

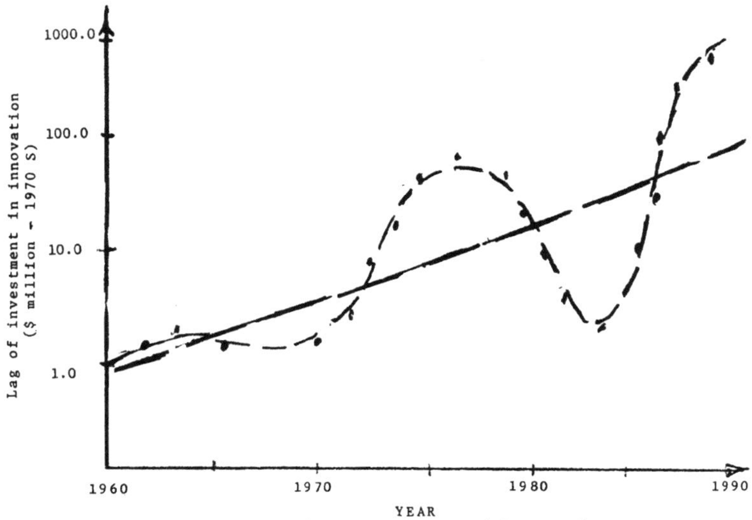

Figure 9.2 Innovation Investment Cycles in Shipyard Automation.

Innovation can be performed solely for the internal purposes of the sponsor to improve his own products, processes, or services directly or indirectly. It may also be performed to develop products, processes, and services which can be sold to others for use, further development, or resale.

Diffusion of technology at any stage of innovation can be internal, external, and international. It can be diffused within a group or company, within an industry or market, or worldwide. The rate of diffusion in these different areas is influenced by the sponsor's strategy, his financial and marketing needs, the cultural and economic environment in which the innovation is performed and, most importantly, the competitive factors. The rate of adoptions to potential adopters is often used as a measure of diffusion, though this approach tends to be a crude approximation because of the difficulty in estimating these two quantities at any stage of the innovation process.

9.1.1 *Commercialization of Innovations.* While it is evident that early diffusion of innovations has many advantages, there are commercial obstacles to such an approach, particularly in countries such as the U.S. whose antitrust laws interfere with joint or even some parallel development of technology, and in which business is done in a fiercely competitive environment. This differs markedly from the approach taken by countries such as Japan, where the government encourages (and sometimes helps organizationally and financially) companies to cooperate in the innovation process until all the basic technological issues are resolved, when cooperating companies each start to use the results of the cooperative innovation process to further their own interest in terms of product, process, and service development.

This approach has allowed Japanese companies to advance technologically at a much more rapid rate than their U.S. competitors and bring better, higher quality, and often more advanced technology to the market at very competitive costs.

In fact it is becoming increasingly evident that rampant competition in technology innovation, with many firms engaged in the same innovation or technology development, has become too expensive for both individual U.S. firms and the U.S. economy as a whole. Recent studies of the computer industry, for example, indicate that over 70% of the cost of innovation by the industry is wasted because other firms (often Japanese companies) came out with more advanced technology before the innovation process is ready with the newly developing technology. Similarly over one-third of the recent bankruptcies of computer hardware companies were the result of wasted R&D and innovation expenditures.

The same appears to happen now in the area of high resolution and other innovative T.V. technologies where the U.S. appears to be well behind Japanese and European developments, largely because of the proliferation of R&D and innovation efforts. Proliferation is also often caused by lack of focus and a lack of technology development policy by the federal government and long-term technology strategies by industry groups and corporations.

While U.S. corporations are short-term profit oriented, they are usually late in their technology diffusion, and often wait until technology under development is near maturity before engaging in wide diffusion. This is largely the result of the perceived domestic competitive environment, yet today the principal competition for both the domestic and foreign markets of U.S. producers are foreign corporations who assume a completely different technology development strategy.

Diffusion of innovation can be generated by the innovator who could be the sponsor of the innovation, the individual or group (organization, etc.) who finances and/or undertakes the innovation process for an invention, or the inventor himself. Diffusion can be actively pursued by use of marketing or simple dissemination of information.

Early diffusion of an innovation provides inflow of capital at a time when the prospects of the innovation are still uncertain. It also encourages adopters who acquire the innovation to use similar technology as the innovator, and not invest in potentially competing technology. On the other hand, it dilutes the potential market of the innovator and unless strict controls are enforced may encourage an adopter of an innovation at an early stage in the innovation process to invest in and undertake accelerated development of the innovation and thereby actually overtake the innovator.

If an innovation is not marketed and diffused early in the innovation process, competitors may be aware of the innovation in general terms, undertake parallel research and development, or acquire competing inventions or innovations. There is also the risk that a long delay in diffusing an innovation may make it obsolete, even before it is fully developed and attains maturity, simply by being overtaken by newer, independently developed technology which benefitted from a faster, more intense innovation process.

9.2 CAPACITY EXPANSION AND COST REDUCTION

Technological change in processes is often aimed at capacity expansion and cost reduction. While it is commonly assumed that capacity and cost reduction are complementary and capacity increase will result in economies of scale, which in turn will reduce costs, this is not necessarily the case.

Increased capacity, unless effectively utilized, may actually increase both fixed and variable costs per unit output, and influence the quality of the output of the operation. In fact, recent industrial experience appears to indicate that improved quality reduces unit costs. Such quality improvements should not be achieved by additional controls, inspections and test, but by design of higher quality in the process, product, and service.

Similarly, it has become evident that more complex products and processes require flexible manufacturing or service approaches and suffer when traditional in-line production or service approaches are used. Flexible manufacturing or service approaches permit discovery of opportunities for product or process innovation and subsequent development. This is the reason that the most significant innovations have been introduced by those who employed such flexible approaches.

In-line mass production and service organizations not only stymie motivation of individuals working on such production or assembly lines, but also prevent individuals from knowing enough about the interaction and interdependence of various processes and tasks (as well as product or service design features) to consider, develop, and suggest improvements or changes which could lead to significant innovations. It was found recently that the number of useful, innovative ideas advanced doubled within one month or change from in-line to flexible manufacturing in a bicycle production plant. Similar advances have been noted elsewhere.

Capacity expansion may not require addition of processes or services but only more effective use of existing technology. Improvements in capacity utilization by load balancing, evening-out of the flow or throughput, effective load sequencing, worker motivation, continuity of employment, and design production/service integration are just a few factors which can increase achievable capacity without sacrificing quality. In fact, quality usually improved and costs were reduced when continuity, effective balancing and sequence of flow of production or service and a high level of utilization is achieved.

Technological change may obviously achieve and is often chosen to increase capacity and/or reduce costs. But this should be as a consequence of technological change. In case additional capacity is needed, all possible improvements and innovations which can be developed in the existing technology used should first be evaluated, to permit determination of an effective "existing technology capability" (capacity, cost, quality, etc.) projection which is then considered as a baseline for comparison with newer/competing technologies. This will prevent costly regrettable mistakes, yet is not always done because many people either become fascinated with newness per se and new technology characteristics or with the often exaggerated and highly speculative projections of new technology capability and performance.

9.2.1 *The Effect of the Size of the Firm.* The escalation in the rate of and scale of investment requirements for innovation in the development of new technology has affected the role of small firms and individual entrepreneurs who traditionally spearheaded technological development. Most inventions and innovations still originate with the entrepreneur and the small firm, but larger organizations, including financial institutions, and government agencies now enter the innovation cycle at a much earlier time, though the latter usually as sponsors.

Most small firms, inventors, and entrepreneurs in new technology are now linked with venture capital firms, other financial institutions, or larger firms who often acquire a controlling interest early in the innovation process of potentially attractive technology and then rapidly take over the management of innovation and subsequent commercialization of the new technology.

Recent surveys of the computer hardware industry in the U.S., for example, indicate that the venture capital and other financial institutions acquired by 1986, nearly 36.8% of the ownership of companies in that field, formed between 1980 and 1985, who attained sales in excess of $10 million by 1986. In the majority of cases, the chief executive and financial officers were appointed or installed by these institutions who, in most cases, also controlled the boards of these companies. Another 10-15% of their share capital was typically owned by larger firms, often firms in the same or related businesses.

The trend of recent mergers and buyouts in the U.S. stock markets is an extension of the finding that it is usually cheaper and faster to acquire new technology or capacity by merger or buyout of a firm, than by internal development. As long as there are a sufficient number of entrepreneurs, inventors, and small firms willing to take the risk of invention and initial innovation, this trend may help to speed up the process of commercialization of new technology.

There is a possibility though that more inventors, entrepreneurs, and small firms may hesitate to take the significant risks involved in the initial R&D and development of technology, if they expect that most of the benefits will escape them, and be reaped by acquirers. This is an important issue which will have to be addressed by policymakers concerned with economic growth which is increasingly dependent on technological advancement. In the future it may be necessary to provide inventors and small innovation firms, particularly in fields of advanced technology, with access to other forms of investment and risk capital or additional protection which goes beyond current laws of ownership and patent rights.

Motivation will remain the single most important force driving technological change. Material benefits are only one of the forces driving motivation for technological change. Others such as status, peer respect, and role in policy and decisionmaking are also important factors. Unless society provides inventors and innovators with such incentives, motivation towards development of technological advance dries up.

9.3 PRODUCT INNOVATION AND DEVELOPMENT

New products originate from basic product ideas, inventions, or discovery. They can be the result of discovery or invention of a new useful object, service, or device. Quite often though new products are developed from recognition of alternate potentials for products in use elsewhere. Examples are the use of laser technology, originally developed as a scientific tool and for instrumentation in printing, or video film technology for still picture taking. As shown in Figure 9.3, the product development process starts with the discovery of basic product ideas or inventions which may be the result of chance invention, directed discovery, or opportunity search.

The product innovation process in which the product idea or concept is refined is often supported by research and development, and preliminary product manufacturing (or if the product is not physical, product structuring) analysis. The innovation process then leads to the product application development which relies on inputs from market research, and regulator requirements (or constraints) to assist product design development, and product processing design. Competition analysis furthermore assists in refining the product application, resulting marketing strategy, and segmentation study, and finally commercialization of the product.

In parallel, product cost analysis based on the product production planning permit an effective pricing strategy to be used in the commercialization stage. Product developments rests on three major parameters:

1. New markets for existing or new products,
2. New materials which make new product development possible or even generate ideas

for new products, and

3. New processes which allow the manufacture of new products.

The basis for new product development is the product idea which tests a new product for newness, relevancy, and usability.

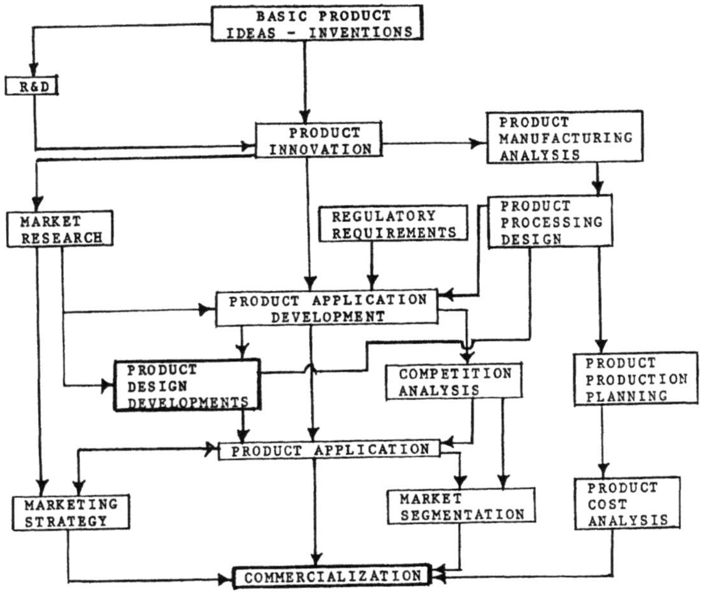

Figure 9.3 Product Development Process.

Products have characteristics, such as form, and attributes. The latter are usually divided into primary attributes and other uses and investigators must always watch out as other uses than the primary use based on the product's principal attributes may rise to dominance.

Product development usually requires entrepreneurial groups who test product potentials, opportunities, and requirements and finally converge on a new product need. Need in turn depends on need recognition, awareness, consciousness, and valuation. Finally, new product developments rely on product use and market forecasting which includes prediction of competitive products and competition in general, market range and market product penetration, and finally product value and potential for new product uses and product market developments.

The innovation of products, on the other hand, is usually affected by processes or services expected or required for manufacture or provision. The availability, condition, and use of processes will often lead to significant product innovation.

Other factors are pressures by competing products, competing producers, and competing markets. Similarly, pricing and price control will have an effect on the product innovation process. With a low market share, a producer is usually a market (and price) follower who will invest in innovation only if his cost (and/or quality) are inadequate, or if such a producer is dissatisfied with his market share and/or profit margin. A market leader, on the other hand, is under perpetual pressure to innovate to maintain his market share, the overall level of demand and to influence the elasticity of demand and therefore profitability.

Pricing of products and resulting effects on market share are linked to product innovation and subsequently process innovation designed to allow efficient manufacture or delivery of the innovated product. Decisions on investment into product innovation, including the rate of investment into innovation, therefore depend on:

1. demand and price elasticity of market for product,
2. dynamics of demand. Time change of demand. Is demand increasing and does the slope of the curve change over time?
3. market share,
4. competitive position,
5. restrictions on increased market share (trade or other regulatory restrictions),
6. financial strength of firm and firm's competitors,
7. cost structure of firm and expected changes in costs,
8. marketing effectiveness,
9. product maturity,
10. product market - Is it an essential, desirable, fashionable, etc. product?

Various decision models exist for the development of an effective pricing strategy to achieve a desired objective such as total profit, market share, etc. These are based on decision tree or game theory algorithms, and assume that there is a fixed (or zero) cost to product acquisition, while production costs are assumed to be constant, functions of output per unit time, and/or time varying. In these models, the product, itself, and the methods of production used in its manufacture (or acquisition) are generally assumed to be invariant over the planning horizon. Yet most products, unless quite mature and close to obsolescence, are subject to innovation and development which, as discussed before, results not only in incremental changes in the product technology but in parallel in the process technology used in its manufacture or supply. As a result, products are subject to the dynamic developments of continual innovation which affects not only their form and capability but also their cost structure as processes (and often materials) are changed in response to changes in the product resulting from the innovation.

9.4 MANAGEMENT OF INNOVATION

To be successful innovation must be planned. It must have specific objectives, schedules, and budgets. Although the outcome of innovation is seldom certain, specific goals in terms of technological improvements which enhance marketability and reduce costs can usually be specified. Such goals based on a firm's objectives then form the basis for the innovation plan.

The goals driving innovation are usually strategic and include consideration of the various factors discussed in the preceding section. They are also quite different from the goals of research, which are usually more diffused and concerned with additions to knowledge or the solution of a problem, without the focus towards commercialization or at least applicability, usefulness, and marketability, which drives goal setting in innovation.

Abernathy and Utterback [1] proposed a dynamic model of innovation, which proposes an observable pattern of change of product innovation, process innovation, and organizational structure. It differentiates between firms that are new to a product area, as compared to those already entrenched. It similarly addresses the development of radical innovation defined as "one which can create new businesses and transform or destroy existing ones". The management of innovation is affected by strategic objectives and operational goals.

Strategic objectives, say the achievement of market leadership, may evolve into various strategic plans and tactics which include such issues as

1. changes in product characteristics,
2. changes in product requirements,
3. establishment and fostering of acceptance of product standards,
4. conditioning of the market and the market environment towards acceptance of firm's new product innovation, and
5. influencing market structure as well as methods used in marketing product.

To facilitate the implementation of such a strategy designed to enhance achievement of the objectives, the firm may have to change its own organizational structure to assure effective interaction among those responsible for the technology innovation, marketing strategy developments, product requirements, product manufacturing or supply, financing of innovation process, and product sales and pricing.

9.5 THE INNOVATION PROCESS

The invention of a new product and its subsequent innovation is usually accompanied by related process innovation. Related processes are often required to permit the new product development, its efficient or economic production, and its effective improvement. While the rate of innovation of a product declines from a peak, after invention if and when a decision to innovate and develop a product is made, until the product reaches maturity, and is usually thoroughly diffused, related process innovation starts slowly as the need for the product, its characteristics, and its production requirements are gradually established. The rate of related process innovation, as a result, reaches a maximum often at the half life of the product innovation period as shown in Figure 9.4, in which we also show the three stages generally associated with product innovation.

1. The <u>initial development stage</u> including technology demonstration which usually follows basic research and discovery or recognition of concept or idea.

2. The <u>product technology demonstration stage</u> during which the technical as well as operational feasibility of the product technology is demonstrated.

3. The <u>product technology development stage</u> during which the product technology is perfected to permit its efficient production. It is during this stage that the related process innovation rate reaches a maximum.

4. <u>Marketing and market introduction stage.</u> In this stage, trial marketing leads to the development of a marketing and diffusion strategy. This may involve licensing of the product technology, including or excluding related process (production) technology. In other words, licensing as part of a diffusion and marketing strategy may involve only product sales or include various stages (or all) of product technology production.

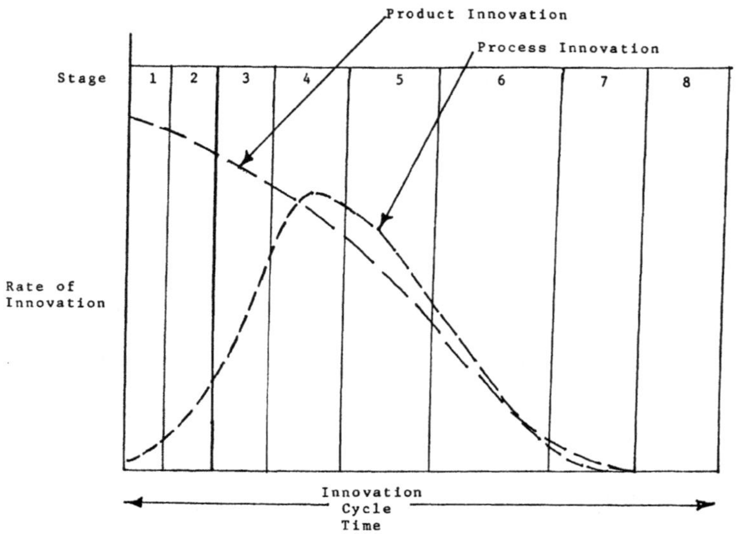

Stage 1 Initial Development
2 Demonstration
3 Development
4 Marketing
5 Market Development
6 Use
7 Maturity
8 Obsolescence

Figure 9.4. The Product and Related Process Innovation Cycle.

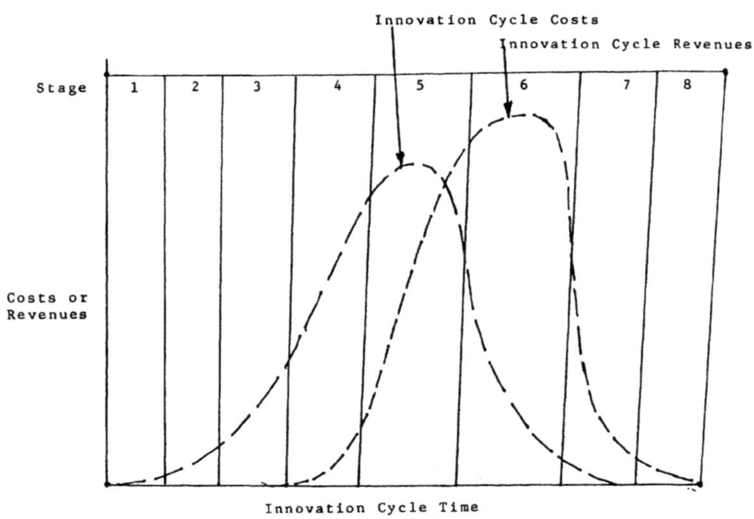

Figure 9.5. Product Innovation Life Cycle Costs and Revenues.

5. Market development stage. During this stage, the diffusion of the product technology (and often its related process technology) takes off, resulting in large revenues.

6. Product technology use stage. Here the technology receives wider and wider acceptance and is widely diffused.

7. Product technology maturity stage. When the product technology stops development because it has reached maturity is largely known widely diffused, no longer a novelty, nor an opportunity for improvement, it is said to have reached maturity which ultimately leads to a serious fall-off in market and diffusion. It also encourages development and introduction of substitute or competing product technologies.

8. Product obsolescence stage. Finally, when the product technology is overtaken by substitute or competing technologies, looses market share, and is subjected to a shrinking diffusion, the technology is said to be in the obsolescence stage.

During the innovation process and its associated life cycle, costs and revenues vary as shown in Figure 9.5.

Expenditures during the basic research and initial development stage are usually quite small, but start to grow rapidly during the product technology demonstration and product technology development stages. They reach a peak during the marketing and market introduction stage, not only because of the large marketing costs usually associated, but also the large related product production process technology development expenditures.

Revenues really only start to be obtained during the market development stage when product and product licensing sales produce income. After reaching a peak, revenues will start to level and then fall off during the product technology use stage during which product and related process innovation costs will start to further decline and then reach zero during the product technology maturity stage. Similarly revenues will continue to fall off during the product technology maturity stage and converge on zero during the product obsolescence stage. The major difficulty in product technology innovation is caused by the phase difference in the cost and revenue flows which require large scale financing, which in turn often forces inventors or innovators to accept less than effective terms.

An important aspect of product innovation cycles is that they will usually be affected by innovation cycles of newer products, trying to catch up or even overtake an earlier product technology. The life cycle of any product technology and its rate of innovation to obsolescence is nowadays increasingly less dependent on decisions based solely on the potentials of a particular product technology but on its potentials - and costs of innovation - relative to those of other often newer product technologies. Development of new product technologies must therefore be undertaken before an existing (often successful) product technology reaches maturity.

References

1. Abernathy, William J. and Utterback, James M., "Patterns of Innovation in Technology", Technology Review, Vol. 80, No. 7, June-July 1978, pp. 40-47.

2. Abernathy, William J. and Townsend, P. L., "Technology, Productivity, and Process Change",

Technological Forecasting and Social Change, Vol. 7, No. 4 (1975), pp. 379-396.

3. Utterback, James M., "Innovation in Industry and the Diffusion of Technology". Science, 183, 1975, pp. 620-626.

4. Rosenbloom, R. S., Technological Innovation in Firms and Industries: An Assessment of the STart of the Art. Working paper, HBS 74-8, Harvard Business School, Boston, 1974.

5. White, George R., "Management Criteria for Effective Innovation", Technology Review, 1978.

6. Burns, T. and Stalker, G. M., "The Management of Innovation", Tavistock, London, 1961.

7. Tilton, J. E., International Diffusion of Technology: The Case of Semiconductors, The Brookings Institution, Washington, DC, 1971.

8. Rosegger, G., "The Economics of Production and Innovation", Pergamon Press, New York, 1986.

10. Decision Theory and Hierarchial Processes

Management of technology deals with decisions of choice, timing, method of approach to technological change, rate of technological change, and commitment or investment in the various stages of technological development and ultimate change. Most of these decisions must be made under conditions of uncertainty, with insufficient and often unreliable information and with only vague knowledge of risk involved and opportunities available. Therefore, success in technological change is often uncertain.

At the same time not all the objectives imposed or used by the decisionmakers may not be under their control and they may furthermore be subject to various internally and externally imposed constraints or regulations. Yet there is an extricable drive towards, and need for, technological change which today, more than ever before, determines the success of individuals, firms, and nations.

Recent history provides some important lessons which show that management of technological change is necessary for effective economic development and growth. While all decisions under uncertainty involve risk, the risk of technological change can be reduced by effective use of all information bearing on the problem, however vague or uncertain. Similarly, it is important to consider all decision alternatives, including the do-nothing alternative, as well as decision alternatives which seem to apply only remotely to the solution of the problem. This is important because human logic often misses good solutions because of bias, inaccurate data, or faulty assumptions.

Decision alternatives usually have expected outcomes which may affect subsequent or conditional decisions (decisions conditioned by the outcomes of the precious decisions). A typical simplified technological change decision problem is shown in Figure 10.1 which describes a simple two-stage technological change decision problem for a production process.

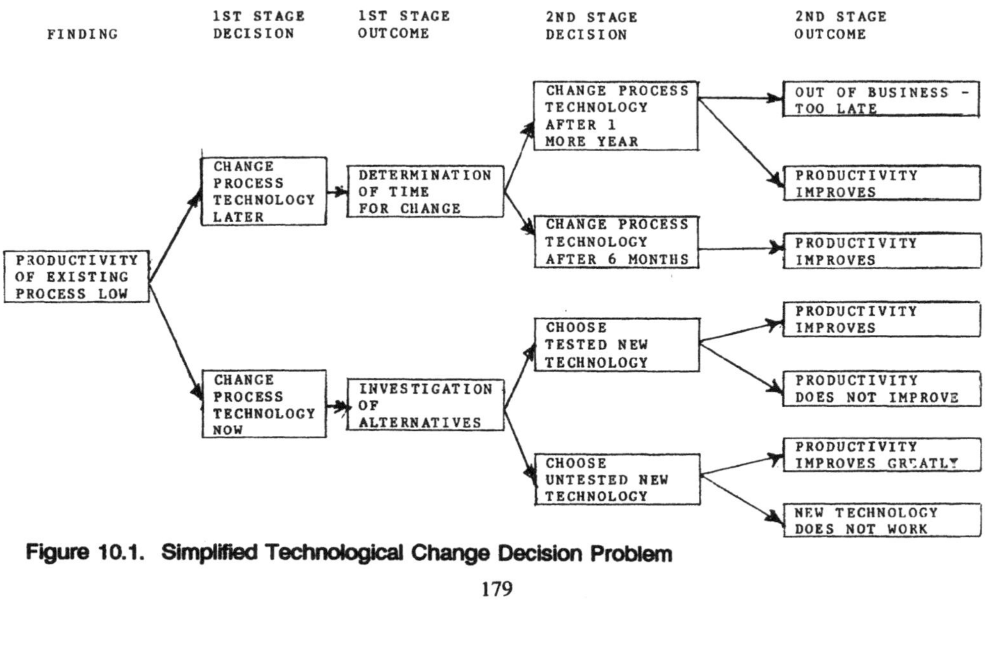

Figure 10.1. Simplified Technological Change Decision Problem

Such a decision problem may consist of many sequential stages of decisions where decisions at each stage may have one or more possible outcomes. These outcomes usually have conditional probabilities of occurrence (conditioned on prior decisions). The number of different outcomes from a decision depend on possible uncontrollable future events (such as the reaction of competitors, actions of government regulators, chance discoveries, and more) and/or known or assumed probability distributions of the outcome.

A typical decision matrix, representing one stage of such a decision process, is shown in Table 10.1 where O^k_{ij} represents outcome j when decision i is undertaken at stage k. For some decisions complete information may be available and the outcome perfectly predictable. However, most technological change decisions are made under conditions of uncertainty. Outcomes O^k_{ij} therefore may have an associated probability p^k_{ij} as shown in the decision matrix in Table 10.2.

	1	2	3
Decision 1	0^k_{11}	0^k_{12}	0^k_{13}
Decision 2	0^k_{21}	0^k_{22}	0^k_{23}
Decision 3	0^k_{31}	0^k_{32}	0^k_{33}

Table 10.1. Decision Matrix - Uncontrollable Future Events.

	E_1	E_2	E_3
Decision D_1	P^k_{11}, 0^k_{11}	P^k_{12}, 0^k_{12}	P^k_{13}, 0^k_{13}
Decision D_2	P^k_{21}, 0^k_{21}	P^k_{22}, 0^k_{22}	P^k_{23}, 0^k_{23}
Decision D_3	P^k_{31}, 0^k_{31}	P^k_{32}, 0^k_{32}	P^k_{33}, 0^k_{32}

Table 10.2. Uncertain Outcome Decision Matrix - Uncontrollable Future Events.

This can also be represented by a decision tree (Figure 10.2).

Let us assume that a production manager has a choice of producing a new product or not. If he produces the new product then there are three possible outcomes:

1. There is a 50% probability that he cannot sell it, with a resulting loss of $70,000.

2. There is a 30% probability of bare acceptance of the product, with a total profit of $50,000.

3. There is a 20% probability of a runaway demand, with a resulting profit of $200,000.

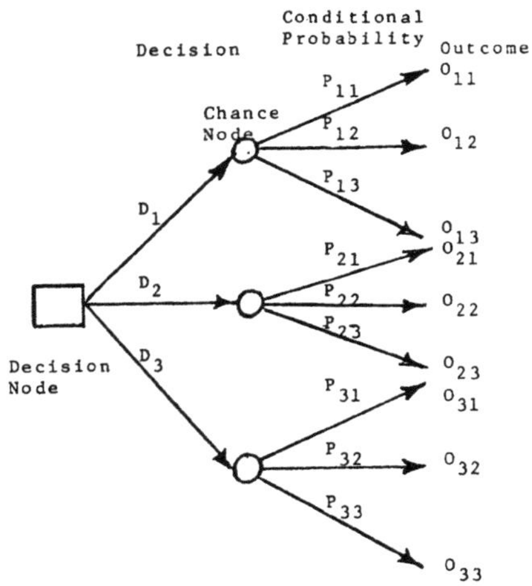

Figure 10.2. One-Stage Decision Tree

Obviously if he decides not to introduce the new product, then there are neither costs nor profits. The expected monetary outcome (value) of the decision to produce the new product is

EMV(Yes) = -0.5 x \$70,000 + 0.3 x \$50,000 + 0.2 x \$200,000 = \$20,000

The expected monetary value of no new product EMV(No) = \$0. As a result, the better decision is obviously to introduce the new product.

Suppose the production manager has an opportunity to improve his knowledge of the potential market for the new product by a market study which costs \$30,000 and will give an indication of the potential popularity of the new product in terms of narrow, wide, or general market popularity. If D_i is the demand for the new product and M_j the market popularity, then we are interested in the conditional probability

$$P(D_i/M_j) = P(D_i, M_j) / P(M_j)$$

where $P(D_i, M_j)$ is the joint probability of demand i and market popularity M_j. Table 10.3 shows these joint probabilities and the marginal probabilities $P(M_j)$ which are the sums of the column of joint probabilities.

Using these results we can now compute the conditional probabilities $P(D_i/M_j)$ as shown in Table 4 and in the resulting decision tree in Figure 10.3.

We can now compute the EMV of the first-stage decision if to introduce a new product without a market study, not to introduce it and no market study, or performance of the market study before making the new product decision. We already know the EMV for the first two decisions 1 and 2.

	M_1	M_2	M_3	$P(D_i)$
D_1	0.30	0.15	0.05	0.50
D_2	0.06	0.10	0.14	0.30
D_3	0.04	0.05	0.11	0.20
$P(M_j)$	0.40	0.30	0.30	1.00

Table 10.3. Joint and Marginal Probabilities

	M_1	M_2	M_3
D_1	0.75	0.50	0.17
D_2	0.15	0.33	0.47
D_3	0.10	0.17	0.37

Table 10.4. Conditional Probabilities

To compute the EMV for decision 3 we first compute the EMV of the second stage decision based on the first stage outcomes M_1, M_2, M_3, and then follow these back to obtain the EMV for decision 3.

$$EMV(M_1) = Max\ [(-0.75 \times \$70{,}000 + 0.15 \times \$50{,}000 + 0.1 \times \$200{,}000);\ 0] = -\ \$25{,}000$$

$$EMV(M_2) = Max\ [(-0.5 \times \$70{,}000 + 0.33 \times \$50{,}000 + 0.17 \times \$20{,}000);\ 0] = -\ \$15{,}100$$

$$EMV(M_3) = Max\ [(-0.17 \times \$70{,}000 + 0.47 \times \$50{,}000 + 0.37 \times \$200{,}000);\ 0] = \$85{,}600$$

The EMV of decision 3 is therefore EMV (Market Study) = -\$30,000 + 0.4 x -\$25,000 + 0.3 x -\$15,100 + 0.3 x \$85,600 = -\$18,850. It therefore does not pay to perform the market study, but go ahead with the production of the product without it.

Outcomes from technological change decisions have different utilities for decisionmakers. If, for example, a change in a process results in a large initial cost increase too large to be sustained, then this outcome may have little utility even though the new process offers huge savings later in time. The utility of early cost savings may be much greater than that of later cost savings, particularly if lack of such saving, or worse each cost increases drive the firm out of business, for example.

A utility function represents the relative value of outcomes of decisions over the whole range of possible outcomes to an individual firm or government. The utility function of a particular decision maker can be obtained by pairwise comparison of the value of outcomes, or a more formal lottery valuation.

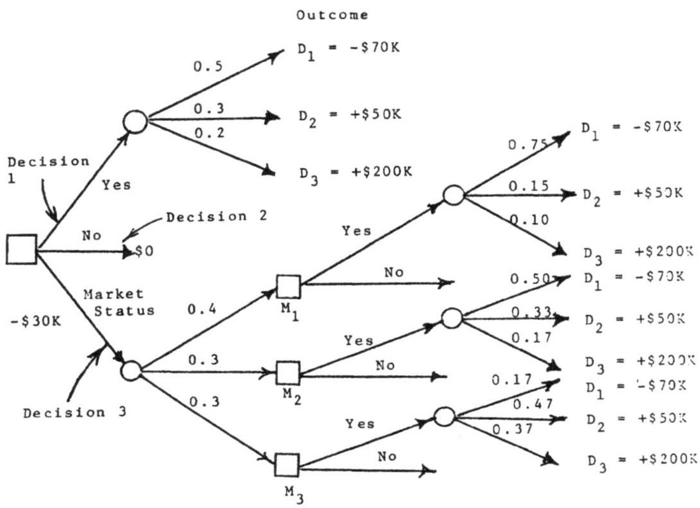

Figure 10.3. Market Research Decision Problem

Assume we want to determine the utility function of a decision maker who confronts a decision with outcomes between -$100 and +$1000. Consider two alternative decisions, A and B, where A has a 50% probability of an outcome with a gain of $1,000, and a 50% probability of an outcome with a loss of $100, while B on the other hand has a certain outcome of x. If a loss of $100 is the maximum loss and a gain of $1,000 is the maximum gain achievable as a result of the decision, as we assumed, we could associate a utility of 1 with the gain of $1,000 and a utility of 0 with the loss of $100. Next, we could ask the decision maker what the value of x has to be to make decision B as good as decision A. If the answer is $200, then we can say that the utility of $200, the outcome of decision B is equal to the utility of the expected outcome of A or we can write:

$$U(200) = p[(U(-100)]U(-100) + p[U(1000)]U(1000)$$

$$= 0.5 \times 0 + 0.5 \times 1 = 0.5$$

Continuing this process, we could define a decision C with a 50% probability of a $1,000 outcome and a 50% probability of a $200 outcome. If the decision maker estimates that the utility of a decision D with a certain outcome of $500 is equal in utility to the decision C described above then:

$$U(500) = p[U(200)]U(200) = p[U(1000)]U(1000)$$

$$= 0.5 \times 0.5 + 0.5 \times 1.0$$

$$= 0.75$$

The process can be continued until a sufficient number of utility points are determined so that a utility function can be drawn as shown in Figure 10.4. The resulting convex utility curve is a sign of risk aversiveness. If the utility function connecting the worst and best possible outcome where to be a straight line, then the decision maker is risk neutral. Conversely a concave utility function would indicate risk proneness of the decision maker.

10.1 ALTERNATIVE PROCESS TECHNOLOGY CHOICE DECISIONS

If there is a choice between alternative process technologies, where one may be a technology in use, a learning or experience curve model is often found useful. Assume Q_1 have been produced by technology 1 and there is a need to produce another q units of output (before

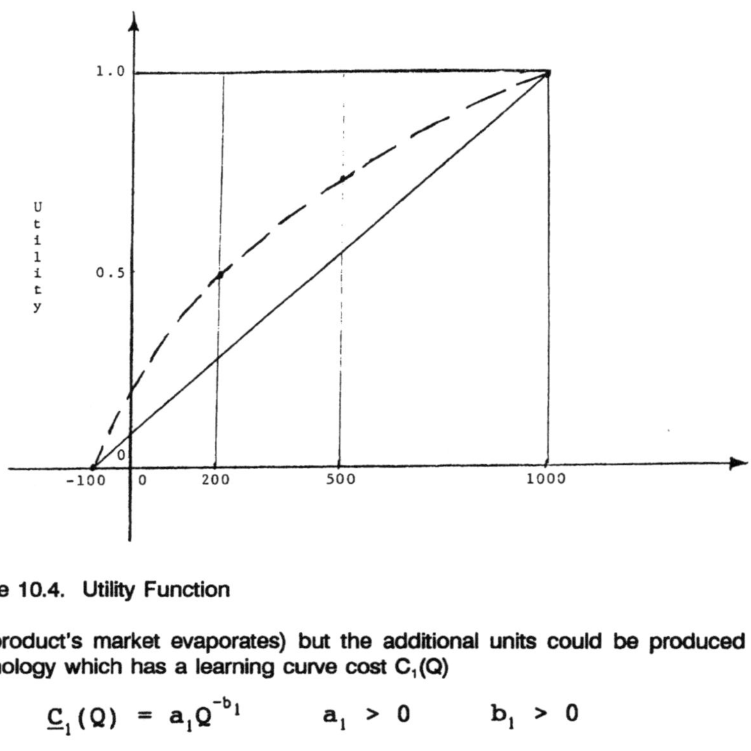

Figure 10.4. Utility Function

the product's market evaporates) but the additional units could be produced by the same technology which has a learning curve cost $C_1(Q)$

$$\underline{C}_1(Q) = a_1 Q^{-b_1} \qquad a_1 > 0 \qquad b_1 > 0$$

or by another technology 2 which has a learning curve cost $C_2(Q)$

$$C_2(Q) = a_2 Q^{-b_2} \qquad a_2 > 0 \qquad b_2 = 0$$

The cost of producing q more units using technology 1 is

$$C(Q_1, q) = \int_{Q_1}^{Q_1+q} a_1 Q^{-b_1} \, dQ$$

$$= \frac{a_1}{1-b_1} [(Q_1+q)^{1-b_1} - Q_1^{1b_1}]$$

$$= (\frac{a_1}{1-b_1}) Q_1^{1-b_1} ((1 + \frac{q}{Q_1})^{1-}$$

If q < < Q, or for

$$q/Q_1 \rightarrow 0, \text{ then } (1 + \frac{q}{Q_1})^{1-b_1} \approx 1 \text{ and } C(Q_1,q) \approx [\frac{a_1 Q_1^{-b_1}}{(1-b_1)}]q$$

If instead of producing the next q units of output using technology 1 we transfer to technology 2, then the cost of producing these q units is

$$C(0,q) = \int_0^q a_2 Q^{-b_2} dQ = \frac{a_2}{1-b_2} [q^{1-b_2}] = \frac{a_2 q^{1-b_2}}{(1-b_2)}$$

and if and only if

$$\frac{a_1 Q_1^{-b_1} q}{(1-b_1)} < \frac{a_2 q^{1-b_2}}{(1-b_2)}$$

should use of technology 1 be continued.

Let us next assume that we want to rationalize the use of two available technologies 1 and 2. Assume Q_1 and Q_2 are the cumulative outputs achieved by each at t, and that the cost of each follows a learning curve

for technology 1 $\quad C_1(Q) = a_1 Q^{-b_1}$

and

for technology 2 $\quad C_2(Q) = a_2 Q^{-b_2}$

Then if a is the additional output required of which q_1 and q_2 are produced by technology 1 and 2 respectively, then the total cost of producing $q = q_1 + q_2$ more units is

$$TC(q;Q_1,Q_2) = \int_{Q_1}^{q_1+Q_1} C_1(Q)dQ + \int_{C_2}^{q_2+Q_2} C_2(Q)dQ$$

$$= \frac{a_1}{1-b_1}[(Q_1+q_1)^{1-b_1} - Q_1^{1-b_1}] + \frac{a_2}{1-b_2}[(Q_2+q_2)^{1-b_2} - Q_2^{1-b_2}]$$

To minimize TC(q; Q_1, Q_2) consider that

$$C_1(Q_1 + q_1) = a_1(Q_1 + q_1)^{-b_1}$$

and

$$C_2(Q_2 + q_2) = a_2(Q_2 + q_2)^{-b_2}$$

There are three cases of interest.

a. $C_1(Q_1 + q_1) < C_2(Q_2 + q_2)$ when it pays to use technology 1 and the cost saving is $C_2(Q_2 + q_2) - C_1(Q_1 + q_1 + 1) > 0$ when using technology 1 is always superior.

b. $C_1(Q_1 + q_1) > C_2(Q_2 + q_2)$ when technology 2 is always superior.

c. $C_1(Q_1 + q_1) = C_2(Q_2 + q_2)$ in which case either $C_1(Q_1 + q_1) > C_1(Q_1 + q_1 + 1) < C_2(Q_2 + q_2 + 1) < C_2(Q_2 + q_2)$ or $C_1(Q_1 + q_1) > C_1(Q_1 + q_1 + 1) > C_2(Q_2 + q_2 + 1) < C_2(Q_2 + q_2)$ but not both. We must therefore use one or the other technology exclusively and

$$TC(q,Q_1,O_2) = \text{Min } \{\frac{a_1}{1-b_1}[Q_1+q)^{1-b_1} - Q_1^{1-b_1}],$$

$$\frac{a_2}{1-b_2}[(Q_2+q)^{1-b_2} - Q_2{}_{1-b_2}]\} = \text{Min } \{TC_1\ (\bar{q}, Q)\ ,\ TC_2(q,Q_2)\}$$

Considering Figure 10.5, we assume that the learning curves of two alternative technologies cross. In that case there is a cumulative output level Q at which the costs of production of the two technologies are equal.

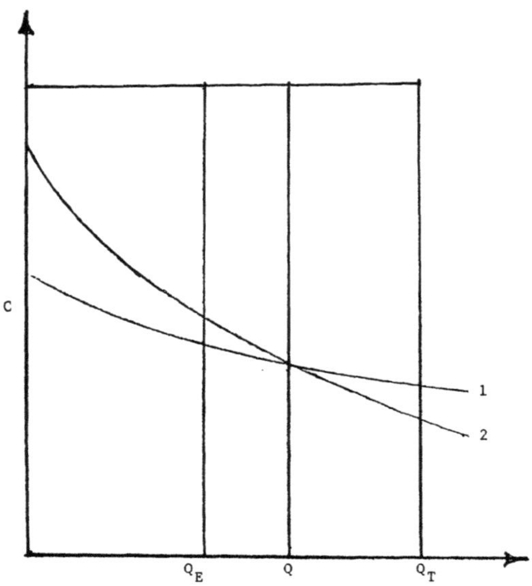

Axes start at Q_1, Q_2 respectively.

Figure 10.5. Learning Curves of Alternative Technologies.

Similarly if Q_T is the total output required then there is a output level Q_E at which the cost of producing the remaining $(Q_T - Q_E)$ units of output is the same for either technology or

$$C(Q_T - Q_E) = \int_{Q_1+Q_E}^{Q_1+Q_t} a_1 \, Q^{-b_1} \, dQ = \int_{Q_2+Q_E}^{Q_2+Q_T} a_2 \, Q^{-b_2} \, dQ$$

and

$$\frac{a_1}{1-b_1} \left[(Q_1+Q_T)^{1-b_1} - (Q_1+Q_E)^{1-b_1} \right] = \frac{a_2}{1-b_1} \left[(Q_2+Q_T)^{1-b_2} - (Q_2+Q_E)^{1-b_2} \right]$$

Therefore under circumstances where both technologies have monotonically decreasing costs, and their costs decline negative exponentially as well as cross with $b_2 > b_1$ (i.e. the learning curve of technology 2 has a steeper slope), then the cost of producing Q_T is lower using technology 1 if $Q_E > 0$ but higher if $Q_E < 0$.

Looking at this problem differently, if Q_1 and Q_2 have been produced at a certain instant in time, there may be an additional output Q_{EQ} which can be produced by both technologies at the same costs, if the learning curves cross as shown. Similarly for any type of learning curves, including non-crossing learning curves, there are values of Q_1 and Q_2 at which the production of another Q_{EQ} units of output costs the same or

$$TC_1(Q_{EQ}, Q_1) = TC_2(Q_{EQ}, Q_2)$$

or

$$\frac{a_1}{1-b_1} \left[(Q_1+Q_{Eq})^{1-b_1} - Q_1^{1-b_1} \right] = \frac{a_2}{1-b_2} \left[(Q_2+Q_{Eq}{}^{1-b_2} - Q_2{}^{1-b_2} \right]$$

the above does not always have a solution, except for the trivial solution $Q_{EQ} = 0$.

10.2 CHOICE AMONG TECHNOLOGICAL ALTERNATIVES

A selection among technological alternatives usually involves many different factors. At the technological level there is the choice among technological alternatives, the timing of adoption, introduction or acquisition of a chosen technology, and the scale as well as rate of introduction of the chosen technology. There are usually a number of measures of performance which affect the selection of technology, each of which has a different bearing on each of the alternative technologies. Finally, there are usually a number of decisionmakers involved in the process of selection of technology. These in turn are influenced by exogenous and endogenous factors, such as government regulation, company finances, the firm's market position, environmental rules, environmental requirements, and more. The typical structure of a problem of managing the choice among technological alternatives is shown in Figure 10.6, and usually consists of decisionmakers who place various weight on different relevant objectives which in turn are affected by exogenous factors such as regulations which influence the achievable measures of performance against which the alternative technologies are measured.

The structure of such a management problem is similar to a decision tree model, with the difference that alternatives or factors at different levels must be ranked or weighted to assure consideration of their respective influence, and the interdependence of the interacting factors at the different levels of the hierarchy must be determined.

Decision tree models consider the outcome of alternative decisions in terms of expected value to the decisionmakers and often apply the utility function of the decisionmakers to rank the desirability of alternative decisions to each in turn. In a hierarchical decision model, we assume knowledge of the relative weight of different actors, such as represented by decisionmakers and objectives at the highest levels of the hierarchy, followed by the relative importance of factors on lower levels always in terms of their relative weight to the actors, objectives, or factors on the next higher level in the hierarchy.

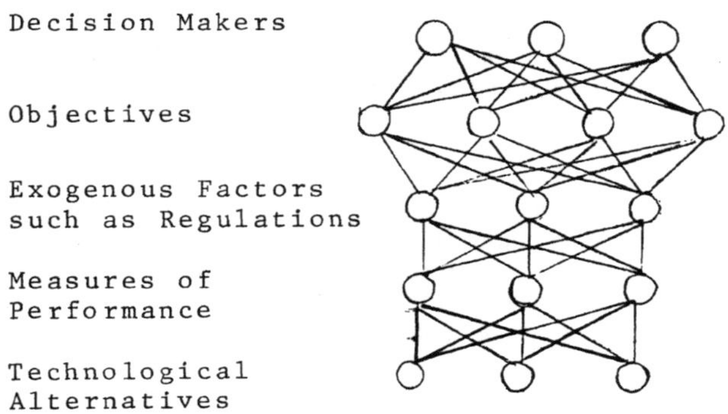

Figure 10.6. Hierarchial Management of Technological Change.

This structural approach to choice making, by priority setting and resource allocation, was developed by Saaty [2] and is called the "Analytic Hierarchy Process" (AHP). The method uses paired comparisons of actors or factors at each level to order priorities at each and sequential levels of the hierarchy. It is particularly applicable to technological choice problems, as these are normally formulated in hierarchical terms, and involve multi-objective and multi-criteria or measure of performance decisions.

Technological choice problems, once formulated and structured as a hierarchy, can be decomposed into levels of factors (which are usually assumed to be independent across their level) and affect elements at the next level. These effects are then measured by paired comparison.

It is interesting to remember that hierarchical structuring, decomposition, and recomposition of problems, are a normal way in which most people present and resolve complex problems. The method really consists of three basic steps:

1. formulation of the hierarchy of the problem;
2. development of the weighing of the factors or elements at each level with regard to contribution of factors at the next level, using weights and paired comparison;
3. setting of the priority weights of the top level actors or factors;
4. subsequent setting of priority weightsd level by level; and,
5. paired comparison composite weights for each.

Problems of technological choice usually involve decisionmakers or actors, objectives, and alternative choices among technologies, their timing, and rate of introduction. The decisionmakers in such problems often plan various roles and have different weights in the decisionmaking process. For example, technological decisions in a small firm may involve technical, financial, marketing, and production managers, each of which has objectives or criteria that he employs in his decisionmaking. Also there is usually some ranking among such decisionmakers. Some, if not all, of the objectives are considered by each decisionmaker, but each will usually put a different weight on a particular objective. Finally, each of the objectives has a different weighing in relation to each of the alternative technological choices.

The above is a typical technological choice decision problem, though quite often several levels of decisionmakers, objectives, and choices may exist and interact. In fact, most decisionmaking involves decisionmakers, criteria (or objectives) and choices. It requires planning, the development of a decision structure, the identification of decisionmakers, objectives, and alternatives, as well as the setting of weights and priorities. It may also involve selection of measures of performance, systen design, and resolution of any conflict in interrelationships or interactions.

This type of three-stage hierarchical decision problem (Figure 10.7) can be solved by:

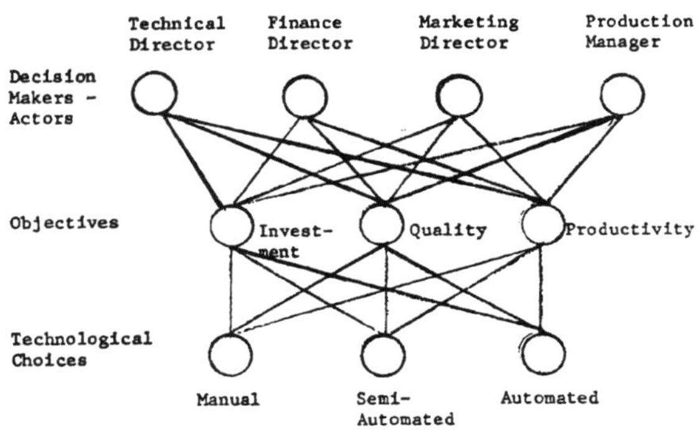

Figure 10.7. Three Stage Model.

- Pairwise comparison of the different actors in terms of their influence or weights on a scale of 100 - say technical/operating director 30%, financial director 20%, marketing director 10%, and production manager 40%.

- Next, the relative importance of the different objectives to each of the actors is compared. In other words, we determine the relative weights of the actor's influence on each objective so as to obtain the relative objective weights.

- Finally, we compare the alternative technologies with respect to their relative contribution to the achievement of each objective and weigh the impact priorities by the weighed objectives.

As shown in Appendix 10C, an eigenvalue approach is used to assure effective weighing of the factors. Similarly methods can be used at discrete time intervals to obtain weighed time varying impact factor estimates. Considering the above three level problem, we would, in addition to the relative weights of the actors or first level weights or priorities,

$$w_1 = [0.3, 0.2, 0.1, 0.4]$$

have the weights each actor places on each of the objectives. For example, the production manager may value productivity 3 times as highly as investment or capital cost and twice as high as quality, while valuing quality twice as important as capital cost. His weighing matrix would then be

	I	Q	P	Row Sums	Eigenvector
Investment I	1	1/2	1/3	1 5/6	.162
Quality Q	2	1	1/2	3 1/2	.309
Productivity P	3	2	1	6	.529

and the eigenvector (corresponding to the maximum eigenvalue) is approximated by:

$$W_{Prod\ max} = (W_{PI}, W_{PQ}, W_{PP}) = (.162, .309, .529).$$

Using similar matrices for each actor, we can now determine the relative importance of each objective by pairwise comparison weighed by the priority weighing of each actor. At the next level weighing matrices are used for each of the alternative technologies with respect to the preceding level factors or objectives. The procedure is then repeated until we obtain the final weight of the alternative technologies in terms of all the interacting weights and interlevel dependencies.

The use of the hierarchical analytical technological choice method requires prediction of the performance and other characteristics of the alternative technologies as well as their availability. Because of the many issues involved, cross-impact forecasting and analysis is considered a useful method for the derivation of the weights and preferences or priorities of the factors of the different levels of the hierarchy.

10.2.1 *Mixed Qualitative and Quantitative Technological Choice Problems.* The major problem in the evaluation of technological alternatives is that the factors of concern in such an analysis are often both qualitative and quantitative. Such mixed factors can be considered separately, considering only one and ignoring the other, considering both in either quantitative or qualitative terms by proper scaling, or integratively as shown in Figure 10.8.

Integrating the results of qualitative and quantitative analysis appears to be the most effective approach. A model designed to integrate qualitative and quantitative factors by producing a unified technology performance measure for each technology alternative under consideration has been proposed by Sharif and Sundararajan [3] and is discussed here. The first step is a technological evaluation which requires:

1. identification and outline specification of the technological alternatives;
2. determination of all qualitative and quantitative factors that influence the choice from among the alternative technologies;
3. classification of all the identified factors which influence the choice of technology;
4. design and formulation of a general model which properly describes the various levels

of qualitative and quantitative factors; and

5. evaluation using the selection model which permits an effective comparison of alternative technologies identified for the defined purpose.

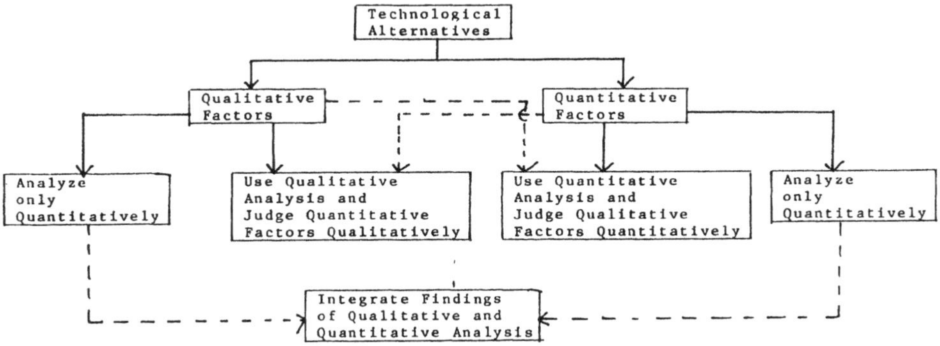

Figure 10.8. Analysis of Technological Alternatives.

Source: Adapted from Reference 6.

Selection from among technological alternatives involves identification of both exogenous (social, political, economic, etc.) and endogenous factors (productivity, technical performance, skill requirements, noise, etc.) that influence the choice decisions.

These factors can usually be identified by technical, operating, management, and other experts, including actual or potential users or buyers of the technology. Once they are identified, it is necessary to divide these factors into (1) qualitative factors and (2) quantitative factors, as discussed above.

We usually find that such a strict classification is inadequate, as there are many qualitative factors which can be quantified even if not on an absolute scale, while there are similarly quantitative factors which have qualitative characteristics which should also be considered. Furthermore factors should usually be grouped in terms of their dominance as well as their type or category. Sharif and Sundararajan [3] suggested three factor classifications:

1. dominant factors which preclude or advance us of a technology,
2. objective or quantifiable factors, and
3. subjective or qualitative factors.

The proposed model integrates the above factors with objective factors measured in some quantitative units (which need not be compatible) and subjective factors in qualitative terms.

The authors suggest use of an index or technology measure for technology i,

$$TM_i = DFM_i \ [\sum_{j=1}^{m} (A_j * OFM_{ij}) + A_0 * SFM_i)$$

where

DFM_i = dominant factor measure for technology i (DFM_i = 0 or 1)
OFM_{ij} = jth objective factor measure for ith technology ($0 \leq OFM_{ij} \leq 1$)
i=1,...,n= number of alternative technologies
j=1,...,m= number of objective factors
A_o, A_j = constants or relative weights with $\sum_{j=0}^{m} A_j = 1$

The dominant factor measure for technology is computed as the product of all the dominant factor indecesl of that technology or

$$DFM_i = \pi_{all\ l} \ [DFI_{il}]$$

where

DFI_{il} = dominant factor index for technology i with respect to factor 1

It is important to note that DFM_i is a zero-one factor and therefore TM_i = 0 if DFM_i = 0.

Considering the objective factors in quantitative terms

$$OFM_{ij} = [OF_i \ / \ \sum_{i=1}^{n} OF_{ij}]$$

where

OF_{ij} = Value of the jth objective factor for i^{th} technology

and

$$\sum_{i=1}^{n} OFM_{ij} = 1$$

Finally to quantify subjective technology factors, the hierarchical analysis approach, developed by Saaty and discussed before, is proposed by the author as follows:

1. As in the AHP, the hierarchy of the problem is formulated by stages of actors, objectives, and alternatives.

2. Each element is weighed at its level in the hierarchy as described in Section 10.2, with respect to the contribution expected to elements or factors at the next level in the hierarchy by means of paired comparison.

3. The procedure is repeated until a composite weight is obtained for each technology alternative which represents the overall measure of that technology with respect to all subjective factors under consideration.

The result, as in AHP, is a normalized eigenvector whose elements are the relative weights of each technology relative to all the subjective factors considered.

The coefficients or weights A_j can be determined by factor analysis as follows:

$$Y_i = \sum_{j=0}^{m} a_{ij} \ F_j + b_i \ U_i$$

where

$i = 1,....n$

F_j = common factor, a function of some unknown variables

a_{ij} = weight for factor F_j and technology i

U_i = a unique factor of technology i

b_i = unique factor weight

and

$$A_i = \frac{a_{i1}}{\sum_{all\ j} a_{ij}}$$

that is the weights are constructed from the first factor loadings.

The procedure is to subject all objective factors OF_j (j = 1,...m) to normalization to obtain OFM_j. Simultaneously subjective factors SF_r (r = 1....k) are subjected to eigenvalue analysis to obtain SFM_i or the subjective factor measure for each of the ith technologies.

The resulting subjective and objective factor measures (SFM and OFM_j's) are multiplied by the dominant factor measure DFM (obtained by multiplying all the dominant factors DF_1's) to obtain the final technology measure TM_i for technology i.

References

1. Raiffa, H., "Decision Analysis", Addison-Wesley, Reading, MA, 1968.

2. Ramanjuan, V. and Saaty, T., "Technological Choice in the Less Developed Countries: An Analytic Hierarchy Approach", Technological Forecasting and Social Change, Vol. 10, 1981.

3. Sharif, M. N. and Sundararajan, V., "A Quantitative Model of the Evaluation of Technological Alternatives", Technological Forecasting and Social Change, Vol. 24, 1983.

4. Barr, D. F., "Decision Making in the Deployment of New Technology: A Practical Approach", European Journal of Operations Research, Vol. 11, 1982.

5. Saaty, T. L., "The Analytic Hierarchy Process - Planning, Priority Setting, Resource Allocation", McGraw-Hill Book Co., New York, 1985.

6. Saaty, T. L. and Vargas, L. G., "Estimating Technological Coefficients by the Analytic Hierarchy Process", Socio-Economic Planning Science, Vol. 13, 1979.

7. Saaty, T. L. and Vargas, L. G., "Hierarchical Analysis of Behavior in Competition: Prediction in Chess", Behavioral Science, 1980.

8. Alexander, J. and Saaty, T., "The Forward and Backward Processes of Conflict Analysis", Behavioral Science, No. 22, 1977.

Appendix 10A - Case Study I - Selection of an Approach and Landing System for Kilimanjaro International Airport (Kia-Tanzania)[1]

INTRODUCTION

Tanzania is located on the east coast of Africa. It is a large country with an area of 945,000 square kilometers and a population of about 20 million concentrated mainly at the periphery of the country. At present, Tanzania has great transportation problems. The existing surface transportation infrastructure is inadequate and in some areas not available.

Air transportation, due to its modal advantage over other systems, is being developed in Tanzania as a mass transportation system catering to passengers, freight, and mail. To serve these needs two international airports (Dar-es-Salaam and Kilimanjaro) and more than 66 other airports and strips, have been developed.

The Tanzania Directorage of Civil Aviation (DCA) is responsible for the planning and development of airports, air traffic control, navigation aids (systems), communication facilities, etc. The DCA's task is particularly hard because surface transportation is quite limited and therefore air transportation is often the only alternative available to the government for the provision of the required rapid and reliable transport services of passengers, freight, etc. which are needed for the country's economic growth.

BACKGROUND

Approach and Landing Systems

In order to ensure that a sound air transport system is in place, DCA continuously plans and introduces improvements, as well as replacements to infrastructure at the various airports. This often involves technological change decisions. In particular, DCA plans the improvement and replacement of navigation aids to enhance the safety and regularity of air transportation.

The present system for the guidance of airplanes landing approach is the Instrument Landing system (ILS) which has been around since 1949 (when it was adopted as a standard system by the International Civil Aviation Organization (ICAO). Although it has undergone continuous technological improvements over time, this technology has reached maturity and its limitations under high density modern jet aviation conditions have become more and more significant. The ICAO has therefore proposed a replacement technology, the Microwave Landing System (MLS).

Due to standardization requirements, the ICAO permits the use of only one new technology at a time. In particular, ICAO usually puts up operational criteria for candidate technologies (as it did in the case of the MLS) and evaluates them before adopting one for universal application.

Technological innovations, since ILS was introduced, have made the use of microwaves a practical possibility. ICAO has since carried out a number of tests on the MLS and so far

[1] Based in part on a case study prepared by Mr. Wilfred O. Malisa, Director, Research and Planning, Directorate of Civil Aviation, Government of Tanzania, while a Fellow at the Center for Advanced Engineering Studies at MIT.

satisfactory signal performance has been achieved. However, MLS landing systems have only been installed in a few countries so far for test purposes. The first of 178 production U.S. MLS ground systems were expected to roll off the production line this spring. (Journal of ATC, July-September 1987).

On the other hand, ICAO has set 1997 as the year when MLS will completely replace ILS. After this date use of ILS will be gradually discontinued. States may not choose to remain with ILS because spares will not be available (as ICAO protection is lifted) and especially because all/most aircraft will then carry MLS receivers on board. The real issue in this case for individual states is therefore not which technology to introduce (as this has already been decided jointly by ICAO forums) but rather when to adopt the new technology.

Replacement of KIA ILS

A brand new ILS was installed at Dar-es-Salaam on October 16, 1984 and so Tanzania still can buy time in deciding when to replace the system at its principal international airport. However, the Kilimanjaro airport ILS is already quite old and is merely surviving on borrowed time. It's performance has become a big problem. It is more off than on the air due to frequent breakdowns. As a result, airplanes have to overfly or divert or cancel flights during bad weather conditions (KIA, apart from being very close to a 20,000 foot high mountain, is known for low clouds during the two rainy seasons of Tanzania - March to July and October to November). This causes not only inconvenience to travelers but also economic losses for airlines (domestic and international) who have to pay route facility (air navigation) fees for very intermittent/unsatisfactory services).

The critical technological questions for Tanzania are as follows:

1. If the ICAO deadline on ILS was not a factor, the decision would have been simple - install another/new ILS.

2. Given that ILS will be phased out worldwide by 1997, should Tanzania go ahead and buy another ILS and get only 10 years of service out of it (ILS useful life is 15-20 years)?

3. When should Tanzania introduce the MLS technology?

ILS TECHNOLOGY

System Description and Operation

As shown in Figure 10A.1, the system comprises three sets of equipments:

1. the localizer transmitter (LOC) which gives guidance in the horizontal plane;
2. the glide path (slope) transmitter which supplies vertical guidance (GP); and,
3. two (sometimes three) marker beacons situated on the approach line which give 'distance to run' to the approaching aircraft.

The three equipments are located as shown and include automatic monitoring and remote control facilities.

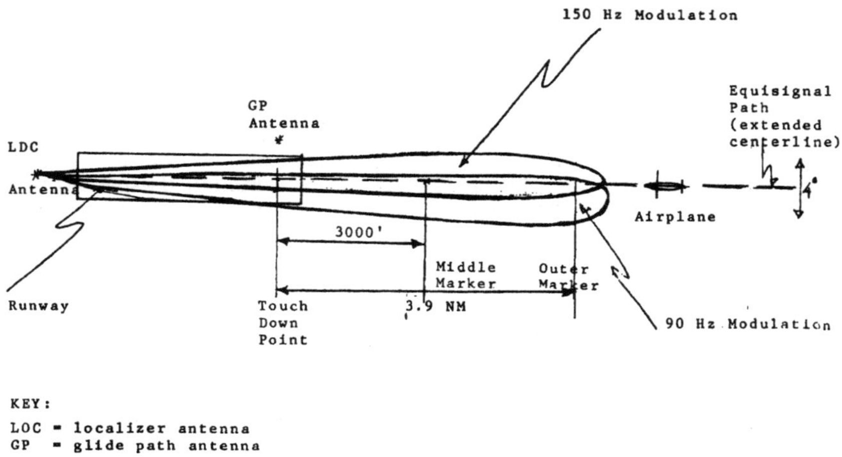

Figure 10A.1. Approach and Landing System

The Localizer (LOC)

The LOC operates on a frequency in the 108.0 MHZ to 112 MHZ band and is required to radiate a radio signal modulated by 90 Hz and 150 Hz tones as shown on the diagram.

On the 'on course line' (equisignal line) the modulation depth of each tone is equal. On either side, the difference in depth of modulation (ddm) between the tones (see overlap area) is proportional to angular displacement (i.e. number of degrees the aircraft is off course) which is picked up by the aircraft's airborne ILS receiver (indicator). As can be seen the region of accurate guidance is only 4 wide.

The Glide Path (GP)

The GP gives the elevation or slope guidance and works in the 328.6 MHZ to 335.0 MHZ band. Again it uses 90 Hz and 150 Hz tones as shown in Figure 10A.2. This arrangement defines an approach path (slope) for the aircraft in the vertical plane of about 3 as shown.

Marker Beacons

The ILS system is completed by 2 (or 3) markers which radiate radio signals on 75 MHZ and situated on the extended centerline of the runway. These radiate vertical "fan" at 90 to the approach line. Each beacon is distinctively coded and its signal can be heard as the aircraft passes overhead.

MLS technology utilizes a time-reference scanning beam technique. The frequency range is 5031.0 MHZ and 5090.7 MHZ. Ground-based equipment transmit position information signals to a receiver in the landing aircraft. The angle measurements are derived by measuring the time difference between successive passes of highly directive narrow fan-shaped beams (please see diagram) and distance measurements are obtained from a suitably located distance measuring equipment (DME).

For each scanning cycle, two pulses are received in the aircraft, the time interval between these being proportional to the angular position of the aircraft with respect to the runway. The elevation function operates in a similar fashion with the beam first scanning upwards and then downwards.

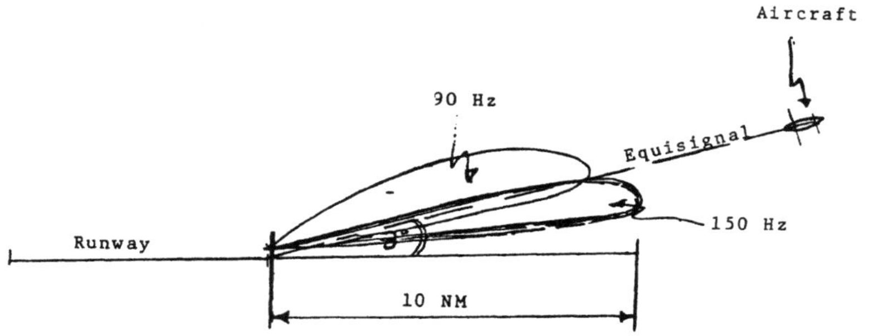

Figure 10A.2 Glide Path

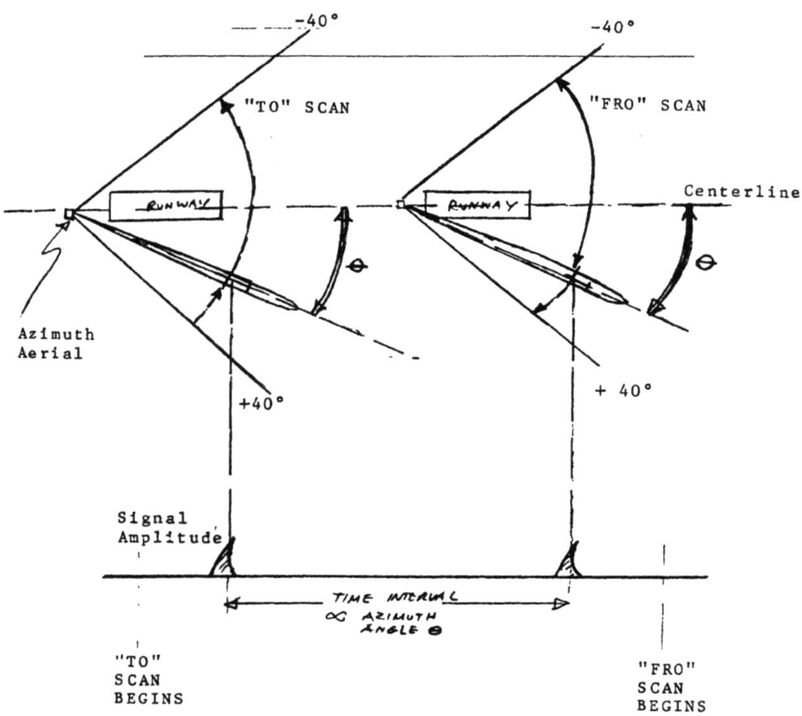

Figure 10A.3 MLS Technology

ILS/MLS PERFORMANCE COMPARISONS

ILS

The ILS has the following shortcomings: (a) approaches are confined to a single narrow path; (b) the number of channels is limited; and (c) the quality of guidance signals is dependent on the nature of the terrain. Hence, siting of ILS is difficult and expensive (a lot of earth works involved). However, regardless of its limitations, that led to development of MLS, present day ILS represents a worldwide, well-established, and reliable system offering safe and operational capabilities. Also large investments have been and will be made during coming years in both ground and airborne equipment. Such states will need time to allow amortization of these investments (at least in part).

MLS

The MLS technique is free from the technical problems associated with ILS. The possible coverage of ± 40 and more will result in increased runway capacity. It is also flexible and much easier to install. Flexibility in particular makes it possible for the better use of terminal airspace as it will allow aircraft to fly curved approaches for optimum management and maximum runway utilization.

Use of MLS will reduce noise impacts near airports and also allow vortex avoidance through curved approach and landing procedures. Despite MLS's great potential users/possib-le adopters are currently concerned with the costs of acquisition and installation and the training of technicians.

TIMING OF INTRODUCTION OF MLS AT KIA

The situation facing Tanzania on the above issue may be structured in the form of a decision tree, in which probabilities may be assigned to the chance branches as shown in Figure 10A.4.

Assumptions on the 4 possible decisions

1. That if we buy ILS now it will surely work well and obtain the benefits. In estimating the EMV (expected monetary value), a life of 15 years has been assumed. Also it was assumed that the ILS will be used only for 10 years, whereafter MLS will be introduced.

2. That if we buy MLS now, there is still a chance as shown that it will not work well. This is new technology and has not had extensive use.

3. Wait 5 years. It is assumed that we should by then have cleared any problems that may arise through redesign and/or modification.

4. Wait 10 years. It is obvious that we will have to buy an MLS which again by then is free from problems.

5. Also present shelf cost of ILS is 30,000 (excluding installation and spares package). This is a very low price indeed as in the electronic industry, due to mass production, costs quickly go down and hence items become cheaper.

The expected cost of MLS on the other hand is 500,000, a cost which is expected to subsequently go down as and when it is mass produced. A thorough estimation of NPVs (EMVs) will require accurate cash flows, inflation rates, etc. Cash flows are not easy to estimate as the technology is still new.

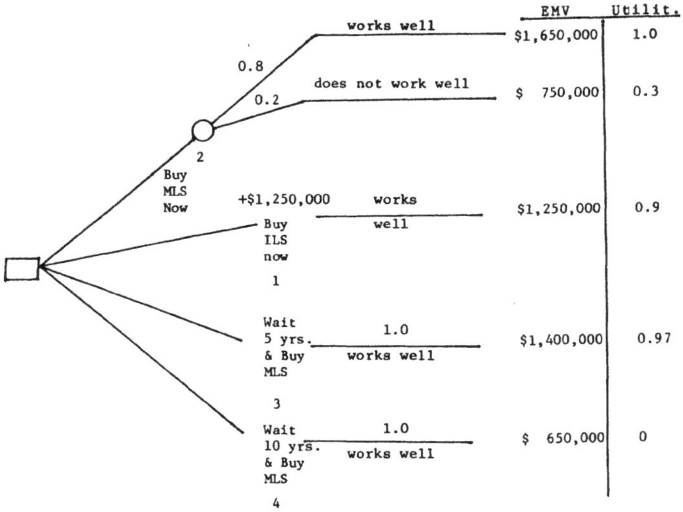

	ILS Current	MLS Current	MLS 5 years	MLS 10 years
Costs $	- 50,000	- 600,000	- 400,000	- 200,000
Benefits $	1,250,000	2,250,000	1,800,000	1,000,000
NOV/Outcome	1,250,000	1,650,000	1,400,000	650,000

Figure 10A.4 Decision Tree

Using the above indicative figures and "folding" the tree back we see that the best choice for Tanzania is to buy an MLS now (option 2) with EMV +#2,070,000 (maximum). However, if we assume the government (DCA) is moderately risk averse as shown in the utility curve presented in Figure 10A.5, we note that our option 2 selection is no longer valid and instead option 3 (wait 5 years) is the preferred alternative with a utility 0.975 (on folding back the tree).

At this point we have established some possible courses of action, which we can further analyze, particularly the preferences of other important decision makers. Use of the analytic hierarchy process to bring other key parties into the decision process is next used.

Figure 10A.6 shows the main actors in the choice of aviation technology in Tanzania. In Figure 10A.7, we show a proposed representation of these various actors and their criteria with regard to the alternatives to be selected.

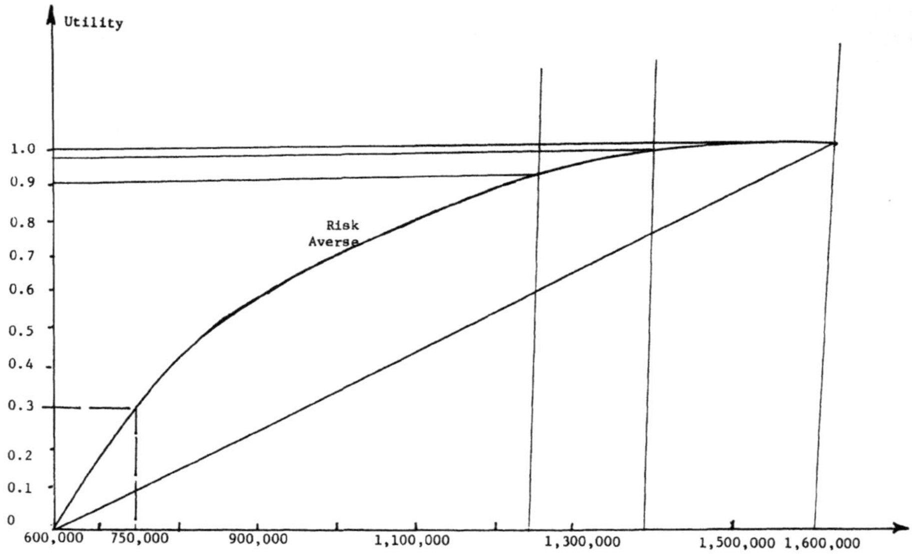

Figure 10A.5 Utility Function

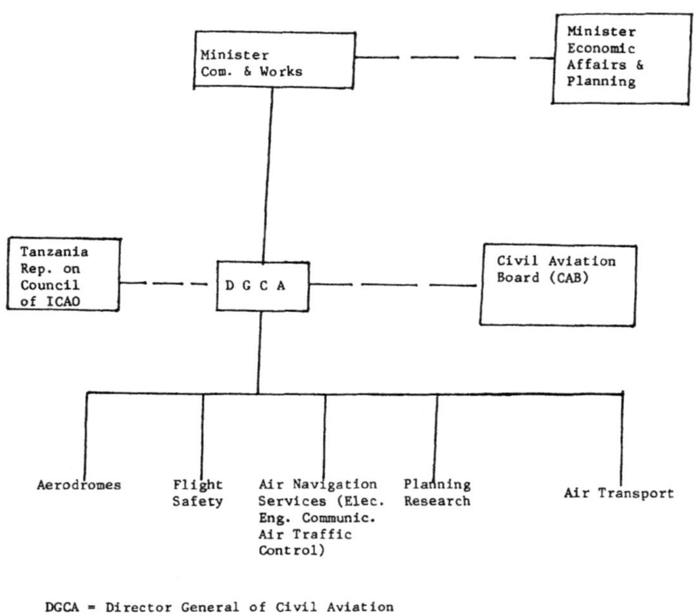

Figure 10A.6 Choice of Technology - Main Actors

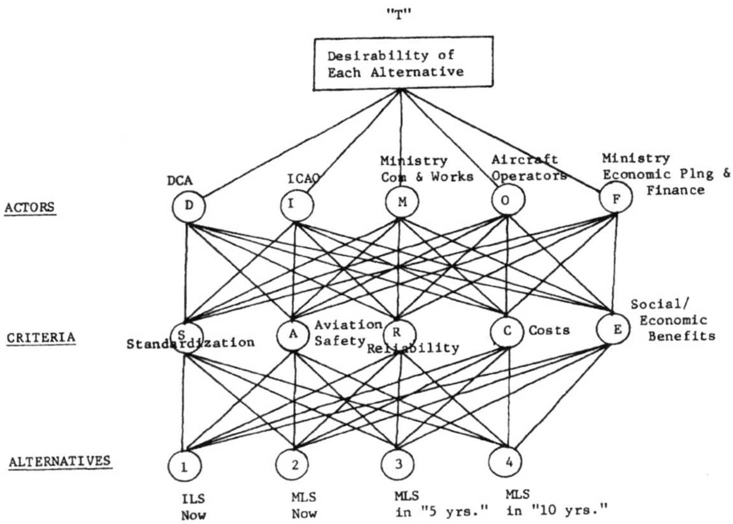

Figure 10A.7 Hierarchial Representation of Factors in the Choice of Timing of Technology (KIA Tanzania)

There are 4 major actors who influence the final choice. The project proposal has to be formulated in DCA (D) and this involves inputs from ICAO (I) and aircraft operators through the civil aviation board.

The project approval process is then negotiated upwards (to the Minister of Communications and Works (M) to consider sectoral/intermodal issues and finally approval, while funding must be obtained from the Minster of Economic Affairs and Planning (Finance).

The objectives considered in this decision process are as shown in Figure 10A.7, standardization (S); aviation safety (A); reliability (R) (and this includes accuracy, maintenance, and flexibility considerations); costs (C); and social/economic benefits (these include efficiency, environment, etc.). The figure therefore shows the hierarchial decision network of all the links in the process. In order to determine which of the alternatives should be chosen for KIA, the following AHP computations are made.

1. First Level

The weights of the main actors in a 1 to 9 scale of importance are:

$W_D = 7$, $W_I = 3$, $W_M = 1$, $W_A = 4$, $W_F = 5$

These weights are used to construct the pairwise comparison matrix to show priorities among main actors:

	D	I	M	A	F	Eigenvectors (Weights)
D	1	7/3	7	7/4	7/5	0.35 = W_{TD}
I	3/7	1	3	3/4	3/5	0.15 = W_{TI}
M	1/7	1/3	1	1/4	1/5	0.05 = W_{TM}
A	4/7	4/3	4	1	4/5	0.2 = W_{TO}
F	5/7	5/3	5	5/4	1	0.25 = W_{TF}

As explained DCA has the main weight.

2. Second Level

The relative importance which the main actors attach to each of the criteria and corresponding pairwise matrices:

(i) For DCA (D)
$W_S=5$, $W_A=7$, $W_R=7$, $W_C=2$, $W_E=4$

	S	A	R	C	E	Eigenvectors (Weights)
S	1	7/5	5/7	5/2	5/4	0.20 = W_{DS}
A	7/5	1	1	7/2	7/4	0.28 = W_{DA}
R	7/5	1	1	7/2	7/4	0.28 = W_{DR}
C	2/5	2/7	2/7	1	1/2	0.08 = W_{DC}
E	4/5	4/7	4/7	2	1	0.16 = W_{DE}

(ii) For ICAO (I)

$W_S=7$, $W_A=7$, $W_R=7$, $W_C=1$, $W_E=1$

	S	A	R	C	E	Eigenvectors (Weights)
S	1	1	1	7	7	0.304 = W_{IS}
A	1	1	1	7	7	0.304 = W_{IA}
R	1	1	1	7	7	0.304 = W_{IR}
C	1/7	1/7	1/7	1	1	0.043 = W_{IC}
E	1/7	1/7	1/7	1	1	0.043 = W_{IE}

(iii) For Ministry of Communications and Works (M)

$W_S=1$, $W_A=4$, $W_R=7$, $W_C=3$, $W_E=5$

	S	A	R	C	E	Eigenvectors (Weights)
S	1	1/4	1/7	1/3	1/3	0.05 = W_M
A	4	1	4/7	4/3	4/3	0.200 = W_M
R	7	7/4	1	7/3	7/5	0.350 = W_M
C	3	3/4	3/7	1	3/5	0.150 = W_M
E	5	5/4	5/7	5/3	1	0.250 = W_M

(iv) For Aircraft Operators (O)

$W_S=4$, $W_A=7$, $W_R=7$, $W_C=4$, $W_E=4$

	S	A	R	C	E	Eigenvectors (Weights)
S	1	4/7	4/7	1	1	0.154 = W_D
A	7/4	1	1	7/4	7/4	0.269 = W_{DA}
R	7/4	1	1	7/4	7/4	0.269 = W_{DA}
C	1	4/7	4/7	1	1	0.154 = W_D
E	1	4/7	4/7	1	1	0.154 = W_D

(v) For Ministry of Finance, Economics Affairs and Planning (F)

$W_S=1$, $W_A=3$, $W_R=7$, $W_C=7$, $W_E=7$

	S	A	R	C	E	Eigenvectors (Weights)
S	1	1/3	1/7	1/7	1/7	0.04 = W_{FS}
A	3	1	3/7	3/7	3/7	0.12 = W_{FA}
R	7	7/3	1	1	1	0.28 = W_{FR}
C	7	7/3	1	1	1	0.28 = W_{FC}
E	7	7/3	1	1	1	0.28 = W_{FE}

3. Third Level

The relative importance/weights of the criteria on each of the alternatives and corresponding pairwise matrices between alternatives.

(i) For Standardization (S)
$W_1=1$, $W_2=7$, $W_3=5$, $W_4=3$

	1	2	3	4	Eigenvectors (Weights)
1	1	1/7	1/5	1/3	0.625 = W_{S1}
2	7	1	7/5	7/3	0.4365 = W_{S2}
3	5	5/7	1	5/3	0.3125 = W_{S3}
4	3	3/7	3/5	1	0.1875 = W_{S4}

(ii) For Aviation Safety (A)

$W_1=2$, $W_2=5$, $W_3=3$, $W_4=2$

	1	2	3	4	Eigenvectors (Weights)
1	1	2/5	2/3	1	0.167 = WA_1
2	5/2	1	5/3	5/2	0.417 = W_{A2}
3	3/2	3/5	1	3/1	0.25 = W_{A3}
4	1	2/5	2/3	1	0.167 = W_{A4}

(iii) For Reliability (R)

$W_1=3$, $W_2=4$, $W_3=6$, $W_4=7$

	1	2	3	4	Eigenvectors (Weights)
1	1	3/4	1/2	3/7	0.15 = WR_1
2	4/3	1	2/3	4/7	0.57 = W_{R2}
3	2	6/4	1	6/7	0.3 = W_{R3}
4	7/3	7/4	7.6	1	0.35 = W_{R4}

(iv) For Costs (C)

$W_1=8$, $W_2=2/3=0.662$, $W_3=1$, $W_4=2$

	1	2	3	4	Eigenvectors (Weights)
1	1	24/2	8	4	$0.687 = WC_1$
2	2/24	1	2/3	2/6	$0.057 = W_{C2}$
3	1/8	3/2	1	1/2	$0.088 = W_{C3}$
4	2/8	6/2	2	1	$0.171 = W_{C4}$

(v) For Economic Efficiency (E)

$W_1=1$, $W_2=3$, $W_3=2$, $W_4=1$

	1	2	3	4	Eigenvectors (Weights)
1	1	1/3	1/2	1	$0.687 = WE_1$
2	3	1	3/2	3	$0.057 = W_{E2}$
3	2	2/3	1	2	$0.088 = W_{E3}$
4	1	1/3	1/2	1	$0.171 = W_{E4}$

The results are:

$W_{D1} = 0.1791$ $W_{D2} = 0.434$ $W_{D3} = 0.269$ $W_{D4} = 0.2185$
$W_{I1} = 0.206$ $W_{I2} = 0.620$ $W_{I3} = 0.423$ $W_{I4} = 0.2855$
$W_{M1} = 0.2278$ $W_{M2} = 0.4206$ $W_{M3} = 0.210$ $W_{M4} = 0.2270$
$W_{O1} = 0.2227$ $W_{O2} = 0.4078$ $W_{O3} = 0.141$ $W_{O4} = 0.2161$
$W_{F1} = 0.2969$ $W_{F2} = 0.3632$ $W_{F3} = 0.231$ $W_{F4} = 0.2134$
$W_{T1} = 0.1495$ $W_{T2} = 0.3659$ $W_{T3} = 0.229$ $W_{T4} = 0.1878$

Appendix 10B - Case Study II - Technological Change Decisions in Copies and Products[1]

INTRODUCTION

As a practical matter, technology selection decisions are usually made by use of subjective approaches. Top management decisions are usually made on the basis of an analysis of the performance of the chosen technology in terms of the goals and objectives of the firm which in turn relies on data collected, analyzed, and integrated at earlier stages.

The criteria used in the final top management decision making must by necessity include not only quantitative but also qualitative factors. Until recently, effective ways to integrate such diverse factors into a quantifiable decision were not available.

The Analytic Hierarchy Process used in this case to evaluate technological change in the next generation of copier and printer products is an example of a logical method for decision making when both quantitative and qualitative factors must be considered.

BACKGROUND

Since the basic technology for reproduction was invented by Chester F. Carson in 1938, the basic licenses for the technology called Xerography (Table 10B.1) has been held by Xerox Corporation, until the license expired in 1965. Because of the superiority of Xerography, many firms, especially Japanese camera companies such as Canon Corporation, Ricoh Corporation, and Konika Corporation, had begun to study Xerography before expiration of the license, because of the business opportunities it offered, but also because it threatened their Diazoelectric copier market (Blue Print). As a result, they were able to enter the copier market soon after the expiration of the Xerography license.

Xerox Corporation had faith in its pioneering effort and assumed that customers would continue to buy their products, because of their experience and reputation for quality.

As Japanese camera companies improved the quality of their models, they also began to improve the basic technology of reproduction by modifying Xerographic technology. A new Xerographic technology called N-P process by Canon was one such example.

The copier market has by now in 1989 become very competitive and both the Xerox Corporation and its competitors are doing research concerning new marking technologies to remove some of the fundamental shortcomings of Xerography.

Technology selection for the next generation copiers is important for copier companies and a wrong decision may cause permanent damage to a firm. The decision is very difficult because criteria for the printer technology selection must contain not only technical factors such as print quality, print speed, and resolution, but also environmental factors such as pollution and noise, economic factors such as running cost and direct material cost, and social factors such as High-Tech image, and so on. In other words, the analyst must deal with both quantitative and qualitative factors, and then integrate them into a logical decision model.

---

[1] Based on case study prepared by Mr. Yoskiaki Takahashi, fellow at the MIT Center for Advanced Engineering Studies as part of course work at MIT.

CHARACTERISTICS

- Positive charge applied on a photosensitive drum by a charging device.

- Light image exposed on the drum. Then photosensitive drum looses resistance in accordance with the intensity of the light (actually with the reflection rate of the documents), and a pattern of positive charge can be produced. The charge pattern on the drum is called image because it is invisible.

- The latent image is developed by the negative charged toner as the developing process.

- The developed toner is transferred to the paper by dielectric force at the transfer process.

- The transferred toner is fixed by heat and pressure on the paper at the fusing process.

SHORTCOMINGS

- Environmental dependence. Xerography takes advantage of electrostatic phenomena.

- Low reliability or quick aging of the consumables such as photosensitive drum and developing material.

- Relatively high running cost.

- Pollution such as air pollution by toner (small plastic particle), ozone emission from corona charging device, and electric noise from corona charging device.

- Technological maturity. Xerography is 50 years old.

Table 10B.1. Characteristics/Shortcomings of Xerography Process.

ALTERNATIVE TECHNOLOGIES

In order to overcome the shortcomings of Xerography, the following new marking technologies are under development in many copier companies.

1. <u>Ink Jet</u>

Ink Jet is a new marking technology which enables us to make an image on a paper by ink droplet. The ink drop can be made by an ink generator which consists of a nozzle with a very small diameter hole, a pump which supplies ink to the nozzle and a Piezoelectric device which pushes the ink from the nozzle to the paper in accordance with an electrical signal. That is, if the electrical signal is received at a certain point, a droplet is generated by the drop generator and the generated ink droplet hit the paper marking an ink image on it. If the electrical signal is off at the next point to the previous point, nothing happens. Therefore, the paper remains without ink image. By repeating this cycle, an image pattern can be obtained.

2. Ionography

Ionography is a marking technology which uses a Dielectric material as latent image holder instead of the photo sensitive drum of Xerography. An image pattern of positive charge is made on the dielectric material by a charging device and then developed by the negative charged toner. After the developing process, the toner image can be processed in the same was as Xerography. The charging device consists of a corona charging device, air flow device, and airflow switching device. The corona discharge device produces numerous numbers of positive charged particles (actually they are the ionized Nitrogen). Air flow then carries the charged particles from the charging device toward the dielectric material. Just before the particles reach the material, the switching device works as an electric gate of the flow. In accordance with the electrical signal, the switching device can make a charged pattern on the dielectric material.

3. Thermal Transfer

Thermal transfer is a new marking technology which consists of a heating device and Doner ink film. The heating device consists of many heat spots and their driving units. The Doner ink film is sandwiched between a paper and the heating spots. If a heating spot is heated by the driving unit in accordance with the electrical signal, coated ink of the Doner film evaporates as an ink vapor and makes an ink dot mark on the paper. If a heating spot remains not heated, the coated ink of the Doner film doesn't evaporate and the paper remains without ink dot image. By repeating the above cycle, an image pattern can be obtained.

4. Laser Xerography

Laser Xerography is a slight modification of Xerography. Instead of using the conventional optical lens system, a positive charge pattern can be produced on the photo-sensitive drum by writing the image using a laser beam. After the formation of the latent image on the photo-sensitive drum, the remaining processes are the same as those used in Xerography. Therefore the main point of Laser Xerography is that it deals with an input image as digital information. That is, the image can be enlarged or reduced or even flipped by only electrical data processing without an expensive optical lens system.

5. Mechanical Impact

Mechanical impact had been the only method which could be used as a computer printer by the time when the first Laser printer was invented by Xerox a few years ago. Mechanical impact technology has less print quality and less output speed than the other alternative technologies, but still exists in the market place because of its price attractiveness.

Mechanical devices such as needles or character heads hit an ink ribbon which is placed between a paper and needles or character heads. In accordance with the electrical signal, needles or character heads are hit by a mechanical hummer and the ink of the ink ribbon is transferred from ribbon to the paper.

MODEL DEVELOPMENT

Classification and Identification

The Analytic Hierarchy Process method is applied here to help in the technology selection for the next generation of copiers and printers.

First the copier and printer market is classified into the following market segments:

- low volume or personal market
- medium volume market
- high volume market

The reason for that segmentation is that the key technology of the next generation products used or adopted may be different for each segment of the market. For example, the requirement of print speed of the personal copier will be different from the requirement of the high speed copier for office use. The following major subjective and objective major factors are identified as having importance in the selection of the technology.

Subjective major factors

- strategic enhancement
- strategic factors
- development risk

Objective major factors

- cost
- print quality
- print speed
- reliability
- portability
- safety
- power consumption

These major factors are decomposed into the following actual design factors.

Subjective Factors

- System Enhancement
    - system applicability
    - technical flexibility

- Strategic Factors
    - technological trend
    - high technological image

- Development Risk
    - uncertainty of the technology
    - development cost

<u>Objective Factors</u>

- Cost
    - running cost
    - initial cost

- Print Quality
    - color adaptability
    - resolution
    - half-tone reproduction

- Print Speed
    - process speed
    - first print output time

- Reliability
    - trouble interval
    - maintainability

- Portability
    - size
    - weight

- Safety
    - acoustic noise
    - ozone emission
    - heat emission
    - dust emission

- Power Consumption
    - standby
    - operation

By combining the identified factors, market segment, and the technology alternatives which are mentioned in the previous section, an analytical hierarchy process diagram can be obtained. Figure 10B.1 shows the partial AHP diagram for the technology selection of the next generation products, and contains both subjective and objective factors, as they relate to each other.

In order to get a consistent solution, the subjective factors and the objective factors should be processed and integrated independently until both factors have the same measuring unit.

QUALITATIVE MODEL FOR THE SUBJECTIVE FACTORS

Table 10B.2 shows the pairwise comparisons of System Applicability, a subjective factor, among technological alternatives. From Table 10B.2, the weights for all technological alternatives with respect to the System Applicability can be obtained. The level of consistency is shown in the table as Consistency Index (CI). The CI is small enough to consider the matrix to be consistent.

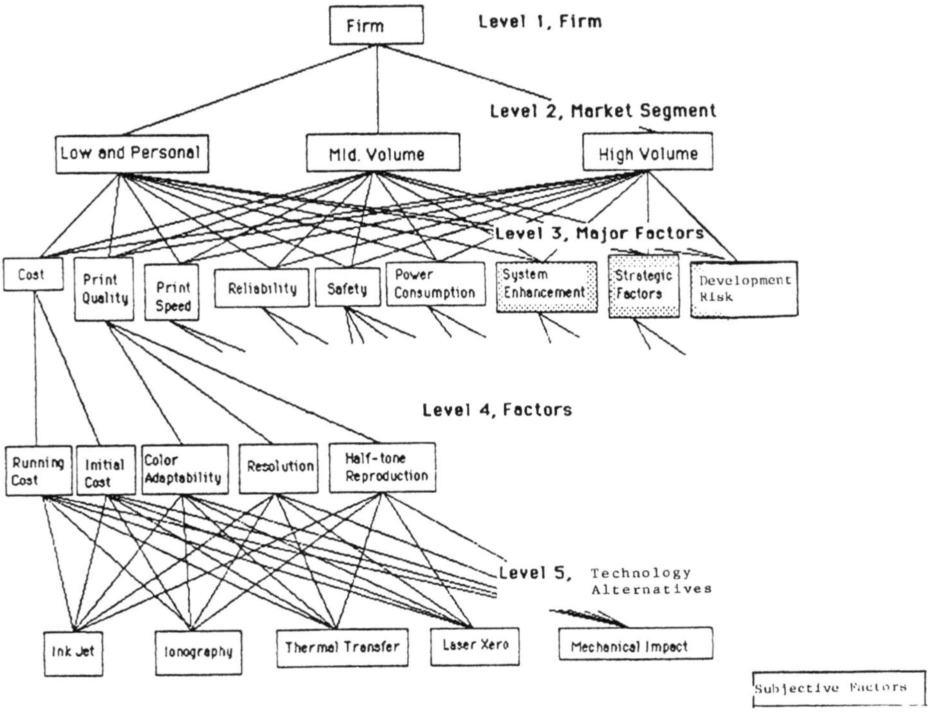

Figure 10B.1. AHP Diagram

	Ink	Ion	Thm	Lar	Meh	Total	Weight
Ink	1	1	3	2	5	12	0.333
Ionograph	1	1	3	2	5	12	0.333
Thermal	1/3	1/3	1	1	2	4.67	0.129
Laser	1/2	1/2	1	1	2	5	0.139
Mech	1/5	1/5	1/2	1/2	1	2.4	0.067
	CI = 0.00075					36.07	1.001

Table 10B.2. Matrix for System Applicability

	Ink	Ion	Thm	Lar	Meh	Total	Weight
Ink	1	2	1	2	3	9	0.302
Ionograph	1/2	1	1/2	1	2	5	0.168
Thermal	1	2	1	1/2	2	6.5	0.218
Laser	1/2	1	2	1	2	6.5	0.216
Mech	1/3	1/2	1/2	1/2	1	2.83	0.095
	CI = 0.055					29.83	1.001

Table 10B.3. Matrix for Technical Flexibility

Similarly, all pairwise comparisons among technology alternatives with respect to the subjective factors are obtained using the same way. Table 10B.3 to 10B.7 show the eigenvect-rs for all technological alternatives under given factors.

	Ink	Ion	Thm	Lar	Meh	Total	Weight
Ink	1	1	1/5	1/3	1/5	2.73	0.067
Ionograph	1	1	1/4	1/3	1/5	2.78	0.069
Thermal	5	4	1	2	1	13	0.321
Laser	3	3	1/2	1	1/2	8	0.197
Mech	5	5	1	2	1	14	0.346
	CI = 0.00025					40.51	1.000

Table 10B.4. Matrix for Uncertainty of Technology

	Ink	Ion	Thm	Lar	Meh	Total	Weight
Ink	1	1	1/8	1/2	1/10	2.72	0.048
Ionograph	1	1	1/8	1/2	1/9	2.72	0.048
Thermal	6	6	1	3	1	21	0.368
Laser	2	2	1/3	1	1/4	5.56	0.098
Mech	10	9	1	4	1	25	0.438
	CI = 0.027					57.04	1.000

Table 10B.5. Matrix for Development Cost

	Ink	Ion	Thm	Lar	Meh	Total	Weight
Ink	1	1	3	2	5	12	0.333
Ionograph	1	1	3	2	5	12	0.333
Thermal	1/3	1/3	1	1	2	4.67	0.129
Laser	1/2	1/2	1	1	2	5	0.139
Mech	1/5	1/5	1/2	1/2	1	2.4	0.067
	CI = 0.00075					36.07	1.001

Table 10B.6. Matrix for Technological Trend

	Ink	Ion	Thm	Lar	Meh	Total	Weight
Ink	1	1	5	2	7	16	0.349
Ionograph	1	1	5	3	7	17	0.371
Thermal	1/5	1/5	2	1/2	2	3.9	0.085
Laser	1/2	1/3	2	1	3	6.83	0.149
Mech	1/7	1/7	1/2	1/3	1	2.12	0.046
	CI = 0.014					45.65	1.000

Table 10B.7. Matrix for High Tech Image

All subjective factors can be integrated into the major subjective factors with the following weights.

- System Enhancement
  - System Applicability (W=0.667)
  - Technical Flexibility (W=0.333)

- Strategic Factors
  - Technological Trend (W=0.667)
  - High-Technology Image (W=0.333)

- Development Risk
  - Uncertainty of Technology (W=0.667)
  - Development Cost (W=0.333)

QUALITATIVE MODEL FOR OBJECTIVE FACTORS

Table 10B.8 shows all the qualitative values for objective factors under the given technological alternatives. The qualitative values of each objective factor may not be normalized by simply summing up the values and by dividing each value by the total value, because each factor may have non-linear utility function. The utility functions for the given objective factors are shown in Figure 10B.2 and have non-linear utility curves.

	Ink	Ion	Thm	Laser	Mech	Unit
Running Cost	2.5	2.8	3.0	2.8	2.3	Cents
Initial Cost	400	1500	280	2000	230	$
Color Adap.	1	1	1	1	0	YesNo
Resolution	300	400	250	400	216	S.P.I.
Half-Tone	0	0	1	0	0	YesNo
Process Speed	10	100	5	60	1	P.P.M.
Fst Output	0	1	0	1	0	YesNo
Trouble Int.	200	18	300	15	80	UMS
Maintenance	30	200	30	200	30	Min.
Size	1200	3000	1000	3500	1800	cm^3
Weight	8	20	5	20	10	Kg
Ac. Noise	5	6	5	6	20	db
Ozone Emission	0	100	0	100	0	--m
Heat Emission	1	200	10	200	1	cal/m
Dust Emission	0	150	0	150	0	P/m
Standby Power	10	300	100	300	20	Watt
Operation P	120	1500	1500	1500	180	Watt

S.P.I. = Spots/inch
P.P.M. = Prints/min
UMS = x 1000 prints/trouble

Table 10B.8. Qualitative Values for the Objective Factors

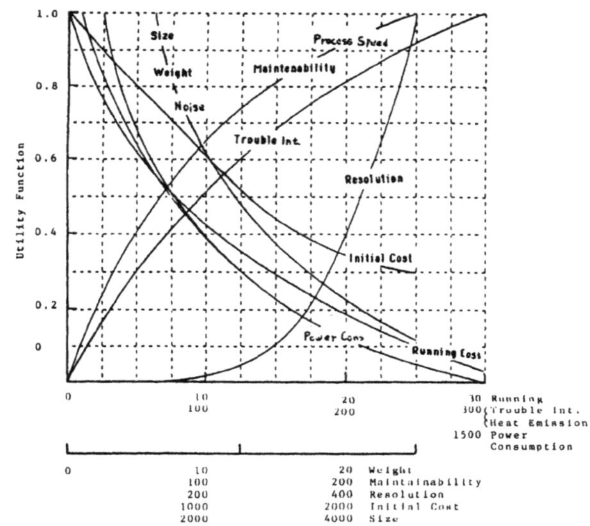

Figure 10B.2. Utility Functions

Table 10B.9 shows the new values of the objective factors, obtained from the values of Table 10B.8 by applying the utility functions mentioned in the previous section.

	Ink	Ion	Thm	Laser	Mech
Running Cost	0.75	0.73	0.04	0.73	0.77
Initial Cost	0.81	0.37	0.87	0.30	0.89
Color Adap.	1	1	1	1	0
Resolution	0.3	1.0	0.13	1.0	0.6
Half-Tone	0	0	1	0	0
Process Speed	0.23	1.0	0.13	0.84	0.03
Fst Output	1	0	1	0	1
Trouble Int.	0.82	0.14	1.0	0.11	0.44
Maintenance	0.26	1.0	0.26	1.0	0.26
Size	0.69	0.25	1.0	0.18	0.53
Weight	0.60	0.16	1.0	0.16	0.47
Ac. Noise	1.0	8.2	1.0	8.2	0.14
Ozone Emission	1	0	1	0	1
Heat Emission	1	0	1	0	1
Dust Emission	1	0	1	0	1
Standby Power	1.0	0	0.39	0	0.89
Operation P	1.0	0	0	0	0.8

Table 10B.9. Modified Qualitative Values for the Objective Factors

Finally, the normalized values for all objective factors with respect to the technological alternatives are obtained as shown in Table 10B.10. Table 10B.10 also shows the values of the subjective factors obtained in the previous section.

At this level, the values of both subjective factors and the objective factors can have a common measuring unit which can be then used to measure all those factors. That means both subjective factors and objective factors can be dealt with equally.

All objective factors can be integrated into the major objective factors with the following weights.

- Cost
    - Running Cost (W=0.333)
    - Initial Cost (W=0.667)

- Print Quality
    - Color Adaptability (W=0.5)
    - Resolution (W=0.3)
    - Half-tone Reproduction (W=0.2)

- Print Speed
    - Process Speed (W=0.9)
    - First Print Output Time (W=0.1)

- Reliability
    - Trouble Interval (W=0.8)
    - Maintainability (W=0.2)

- Portability
    - Size (W=0.9)
    - Weight (W=0.1)

- Safety
    - Acoustic Noise (W=0.667)
    - Ozone Emission (W=0.111)
    - Heat Emission (W=0.111)
    - Dust Emission (W=0.111)

- Power Consumption
    - Standby (W=0.2)
    - Operation (W=0.8)

Table 10B.11 shows the pairwise comparisons among the major factors which include both subjective and objective factors. The table depicts the performance requirement for the low and personal products. So, cost and strategic factors are the most important factors for the market. Conversely, print speed and system enhancement are the least important factors. Similarly, Table 10B.12 shows the results of the medium volume market. the table can be obtained by the same was as Table 10B.11. Table 10B.12 depicts that print quality, reliability, and safety are the most important factors for the market.

	Ink	Ion	Thm	Laser	Mech
Running Cost	0.248	0.242	0.013	0.242	0.255
Initial Cost	0.250	0.144	0.269	0.093	0.275
Color Adap.	0.250	0.250	0.250	0.250	0.000
Resolution	0.099	0.330	0.043	0.330	0.198
Half-Tone	0.000	0.000	1.000	0.000	0.000
Process Speed	0.103	0.448	0.058	0.377	0.014
Fst Output	0.333	0.000	0.333	0.000	0.333
Trouble Int.	0.327	0.056	0.398	0.044	0.175
Maintenance	0.094	0.360	0.094	0.360	0.094
Size	0.260	0.094	0.377	0.068	0.200
Weight	0.251	0.067	0.418	0.067	0.197
Ac. Noise	0.054	0.442	0.054	0.442	0.008
Ozone Emission	0.333	0.000	0.333	0.000	0.333
Heat Emission	0.333	0.000	0.333	0.000	0.333
Dust Emission	0.333	0.000	0.333	0.000	0.333
Standby Power	0.439	0.000	0.171	0.000	0.390
Operation P	0.556	0.000	0.000	0.000	0.444
Applicability	0.333	0.333	0.129	0.139	0.067
Flexibility	0.302	0.168	0.218	0.218	0.095
Tech Trend	0.333	0.333	0.129	0.139	0.067
High Tech	0.349	0.371	0.085	0.149	0.046
Uncertain	0.067	0.069	0.321	0.197	0.346
Dev. Cost	0.048	0.048	0.368	0.098	0.438

C1 = 0.00275

Table 10B.10. Normalized Quantitative and Qualitative Values for the Factors

	Cost	Qual	Sped	Reli	Port	Safe	Power	Enha	Strt	Risk	Weight
Cost	1	6	15	8	2	10	3	12	1	7	0.283
Qual	1/6	1	2	1	1/3	2	1/2	2	1/5	1	0.045
Sped	1/15	1/2	1	1/2	1/8	1/2	1/5	1	1/15	1/2	0.019
Reli	1/8	1	2	1	1/4	1	1/3	2	1/8	1	0.038
Port	1/2	3	8	4	1	5	1	6	1/2	4	0.144
Safe	1/10	1/2	2	1	1/5	1	1/5	1	1/10	1	0.031
Power	1/3	2	5	3	1	5	1	4	1/3	2	0.103
Enha	1/12	1/2	1	1/2	1/6	1	1/4	1	1/12	1/2	0.022
Strt	1	5	15	8	2	10	3	12	1	7	0.279
Risk	1/7	1	2	1	1/4	1	1/2	2	1/7	1	0.035

C1 = 0.00275

Table 10B.11. Matrix for Low and Personal Market Segment

	Cost	Qual	Sped	Reli	Port	Safe	Power	Enha	Strt	Risk	Weight
Cost	1	1/2	15	1/4	4	1/3	2	12	6	3	0.083
Qual	2	1	2	1/3	8	1	4	2	10	5	0.153
Sped	1	1/2	1	1/4	3	1/3	2	1	5	3	0.074
Reli	4	3	4	1	12	1	7	4	20	10	0.285
Port	1/4	1/8	1/3	1/12	1	1/10	1/2	1/4	1	1	0.020
Safe	3	1	3	1	10	1	5	3	15	10	0.225
Power	1/2	1/4	1/2	1/7	2	1/5	1	1/2	3	1	0.039
Enha	1	1/2	1	1/4	4	1/3	2	1	5	3	0.078
Strt	1/6	1/10	1/5	1/20	1	1/15	1/3	1/5	1	1/2	0.016
Risk	1/3	1/5	1/3	1/10	1	1/10	1	1/3	2	1	0.028

C1 = 0.039

Table 10B.12. Matrix for Medium Volume Market Segment

	Cost	Qual	Sped	Reli	Port	Safe	Power	Enha	Strt	Risk	Weight
Cost	1	1/10	1/15	1/10	1/7	1/3	3	1/5	1/12	1	0.019
Qual	10	1	1	1	1	3	10	2	5	10	0.198
Sped	15	1	1	2	2	5	30	3	8	15	0.254
Reli	10	1	1/1	1	1	3	10	2	5	10	0.197
Port	7	1	1/2	1	1	2	20	1	4	7	0.107
Safe	3	1/3	1/5	1/3	1/2	1	10	1	1	3	0.063
Power	1/3	1/30	1/30	1/30	1/20	1/10	1	1/15	1/15	1/3	0.007
Enha	5	1/2	1/3	1/2	1	1	15	1	2	5	0.097
Strt	2	1/5	1/8	1/5	1/4	1	5	1/2	1	2	0.038
Risk	1	1/10	1/15	1/10	1/7	1/3	3	1/5	1/2	1	0.020

C1 = 0.098

Table 10B.13. Matrix for High Volume Market Segment

Table 10B.13 also shows the similar results for the high volume market segment. From Table 10B.13, it becomes clear that print quality, print speed and reliability are the most important factors for the market. Conversely, power consumption and cost are the least important factors.

TECHNOLOGY SELECTION

The combined weights for each technological alternative can be obtained by multiplying the weights obtained in Tables 10B.10 through 10B.13.

Table 10B.14 shows the total weight for each technology alternative with respect to the market segmentation. From Table 10B.14, it becomes clear that ink jet, thermal transfer, and ionography will be the best technology selection for the low volume market, medium volume market and high volume market respectively. If a firm can afford to develop all those technologies at the same time separately, the above decision would be the best. However, if the firm cannot afford this, the decisionmaker must select one (or two) technologies from the three alternatives.

	Low	Mid	High (Market)
Ink Jet	0.2909	0.2248	0.2087
Ionography	0.1935	0.2075	0.2424
Thermal Transfer	0.1916	0.2350	0.2352
Laser Xero	0.1290	0.1892	0.2048
Mechanical Impact	0.1941	0.1445	0.1087

Table 10B.14. Total Weight for Each Technology Alternative

Table 10B.15 shows a pairwise comparison matrix among market segments with respect to the importance of the market for the firm. By multiplying the weights of Table 10B.15 to the weights of Table 10B.14, the final weights for the technology alternatives can be obtained as Table 10B.16.

	High	Mid	Low	Total	Weight
High	1	2	5	8	0.570
Mid	1/2	1	3	4.5	0.321
Low	1/5	1/3	1	1.533	0.109
	C1 = 0.041				

Table 10B.15. Matrix for the Importance of the Market

Ink Jet	0.2228
Ionography	0.2259
Thermal Transfer	0.2304
Laser Xero	0.1915
Mechanical Impact	0.1295

Table 10B.16 Final Weight for Each Technology Alternative

In conclusion, therefore, the firm should select thermal transfer as the first priority technology for the next generation products. If the firm can afford to develop another alternative technology, it should develop either ink jet or ionography as the second priority technology.

References

1. Sharif, M. N. and Sundararajan, V., "A Quantitative Model for the Evaluation of Technological Alternatives", Technological Forecasting and Social Change 24, 15-29, Elsevier Science Publishing Co., Inc., 1983.

2. Ramanujam, Vasudevan and Saaty, Thomas L., Technological Change in the Less Developed Countries, Technological Forecasting and Social Change 19, 81-98, Elsevier Science Publishing Co., Ltd., 1981.

3. Frankel, E. G., 13.651 Lecture Note on Management of Technological Change, MIT 1988 Spring Term.

Appendix 10C - The Analytic Hierarchy Process

The Analytic Hierarchy Process (AHP) was developed by Thomas L. Saaty as an analytical tool for hierarchial planning and resource allocation [Ref. 1]. Many problems in the management of technological change can be structured as hierarchial decision networks, starting with top level decision makers and descending through different levels to the final and basic technological decision alternatives. The network branches represent the relationships or dependencies between decisionmakers, events, objectives, etc. at one and the next level. Decisionmakers, events, objectives, and finally alternatives, have characteristics which affect their standing and attractiveness to those at the next level. In other words, judgement on the relative importance of events, objectives, outcomes, etc. at the next level to decisionmakers, events, objectives, etc., at the next higher level are usually made on the basis of characteristics, factors, and priorities of preferences.

AHP is a formal step-by-step method which insures that such judgements are quantified effectively and in such a way that it allows a quantitative interpretation of the judgment among all the events, objectives, or outcomes at a particular level in the decision hierarchy.

Assume for example that D_{ik} are a set of ij decisions, objectives, outcomes, etc. at a certain k^{th} level in the decision hierarchy and that a_{ik} are pairwise comparisons or valuations or judgements on a pair D_{ik} and D_{jk} by decisions, objectives, or outcomes one level higher in the hierarchy. Hierarchies can be organized in different ways. In a management of technological change decision problem, levels of the decision hierarchy may consist of the following.

Hierarchial Levels	Elements
Top Level Objective	Success of Company
Factors Contributing to Top Level Objective	Profit, Market Share, Sales, etc.
Actors (Decisionmakers)	Production Manager, Sales Manager, Financial Manager, etc.
Actor's Objectives	Quality, Cost, Employment, Product Development, etc.
Technological Process Alternatives	Old Process, New Process, Advanced Process
Technological Product Alternatives	Old Product, New Product, Advanced Product

The above hierarchy could be represented by a diagram or network, in which each element at a level in the hierarchy is connected with each element at the next higher and the next lower level in the hierarchy, if it exists. Considering an element at one level now, which may be an alternative, a decisionmaker, an objective, etc., we now develop a pairwise preference comparison of pairs of elements at the next level in relation to that element. If, for example, the element is the production manager, then a pairwise preference comparison by him would be performed with respect to each of the different production process alternatives.

For example if, using a scale of 1 to 4, the production manager prefers the advanced process (A) to the new process by a factor of 2 and the new process to the existing process (E) by a factor of 2 and the advanced process to the existing process by a factor of 3, then a preference matrix M = (m_{ij}) could be formed as follows

M =

	A	N	E
A	1	2	3
N	1/2	1	2
E	1/3	1/2	1

where the production manager similarly prefers N only half as much as A and so forth.

Once preference matrices are developed for each event, decisionmaker, etc., in relation to all events at the next higher level as just shown, vectors of priorities can be computed from these matrices. This, in mathematical terms, implies the computation of the principal eigenvectors of the matrices. The eigenvector can be computed by several approximate methods in addition to the use of an exact solution. In the approximate methods, we either

a. Divide the elements of each column by the sum of the elements in each column and then add the elements in each resulting row and average by dividing this sum by the number of elements in the row.

b. Multiply the elements in each row and take the nth root, where n is the number of elements in a row. Then normalize the resulting number for each row by dividing by n.

Finally, an exact solution requires raising of the matrix to an arbitrarily large power, when the sum of each row is divided by the sum of the elements in the whole matrix to give the priority vector. Applying the first approximate method to our example we obtain priority vector = [0.538, 0.297, 0.163]. The second method results in a priority vector = [0.549, 0.296, 0.169]. In other words the approximate methods are pretty close and converge on the exact solution.

The use of the eigenvector as a measure of priority or preference can be shown analytically by weighing the preference measures. Considering m_{ijk} as the relative preference of event i over j by k, then m_{ijk} = $1/m_{ijk}$ and the matrix m is reciprocal. Similarly if the matrix M is consistent, then

$$m_{imk} = m_{ijk}\, m_{jmk} \quad \text{for all } i,\ j,\ m,$$

and comparisons are based on exact measurement. This also implies that the weight w_{ik} are known where

$$m_{ijk} = w_{ij}/w_{jk} \qquad i,\ j = 1,2.....$$

and

$$m_{jmk} = w_{jk}/w_{mk} \qquad j,\ m = 1,2.....$$

thus

$$m_{ijk}m_{jmk} = \frac{w_{ik}}{w_{jk}} - \frac{w_{jk}}{w_{mk}}$$

and

$$m_{jik} = 1/m_{ijk}$$

Now

$$\sum_{j=1}^{n} m_{ijk}\ x_{ik} = y_{ik} \qquad i = 1,.....n$$

and

$$m_{ijk}\ w_{jk}/w_{ik} = 1 \qquad i,\ j = 1,.....n$$

then

$$\sum_{j=1}^{n} m_{ijk}\ w_{jk}/w_{ik} = n \qquad i = 1,.....n$$

and

$$\sum_{j=1}^{n} m_{ijk}\ w_{jk} = nw_{ik} \qquad i = 1,.....n$$

which is equivalent in matrix notation to

$$M_w = nw$$

As noted by Saaty [Ref. 1] this formula expresses the fact that w is an eigenvector of M with eigenvalue n or

$$M \quad \begin{array}{c|cc} & M_1 & M_n \\ \hline M_1 & w_{ik}/w_{ik} & w_{ik}/w_{nk} \\ . & . & . \\ . & . & . \\ . & . & . \\ M_n & w_{nk}/w_{ik} & w_{nk}/w_{nk} \end{array} \quad \begin{bmatrix} w_{ik} \\ . \\ . \\ . \\ w_{nk} \end{bmatrix} = n \begin{bmatrix} w_{ik} \\ . \\ . \\ . \\ w_{nk}/w_{nk} \end{bmatrix}$$

If m_{ijk} is not an exact measurement by a subjective judgement and therefore deviates from the exact value of w_{ik}/w_{jk} the equation $m_w = nw$ does not hold.

But if there are numbers λ_i which satisfy the equation

$$M_\lambda = \lambda x$$

where λ is the vector [$\lambda_1 ... \lambda_n$] and therefore are the eigenvalues of matrix M and if the diagonal values m_{ijk} are all one tor all i then

$$\sum_{i=1}^{n} \lambda_i = n \text{ and all eigenvalues}$$

are zero except n, which in the consistent case is therefore the largest eigenvalue of M. In other words we should be interested in finding the vector w which satisfies the equation

$$M_w = \lambda_{max} w$$

and also assure that

$$\sum_{i=1}^{n} w_{ik} = 1$$

Saaty furthermore suggests the use as a contingency index

$$(\lambda_{max} - n) / (n-1)$$

Another issue in AHP is the scale. In our little example we used a scale of 1-4. In problems with larger numbers of events at the various levels and greater potential variability, a larger scale would be useful, yet as shown in Ref. 1, a scale of 1-9 is usually more than adequate and little further accuracy is attained by widening the scale.

It is practice to use 1 as the measure of equal importance and the upper level of the scale, say 9, as the measure of absolute dominance or overriding importance in relation to the compared event. In other words, $m_{ijk} = 1$ means that events i and j have equal importance to the next higher level event k, and $m_{ijk} = 9$ means that event i absolutely dominates event j as far as next higher level event k is concerned.

The weights are usually obtained by a concensus method, such as the Delphi technique. There are obviously many instances where one or a limited number of actors or experts are available, involved or knowledgeable in the weighing judgement or estimation when a unique opinion evaluation must be used.

REFERENCES

1. Saaty, Thomas L., 'The Analytic Hierarchy Process', McGraw Hill Book Company, New York, 1985.

11. The Economic Impact of Technological Change

The economic impacts of technological change are pervasive. Technological change affects all nations on a macro scale and firms as well as individuals on a micro scale.

Early studies of the economic impact of technological change by Solow [1] and others [2, 3, 4] concentrated on the macro effects of output, employment, investment, and other resource use. The aggregate impact of technological change is important and its study explains the effect of national and international economic and industrial policies, international conventions, as well as industry developments. It is not designed to explain the economic impact of the management of technological change at the process, product, or service level. Yet it is often at the micro level where the specific decisions are made on what technology to choose, when to introduce it, and at what rate and to what purpose to apply it.

Earlier studies clearly show the relationships between technological change and economic growth. The U.S. as an example, achieved an increase in the output per manhour or labor productivity of 96.4% between 1950 and 1984, while GNP and the value of industrial production grow by 212% and 296% respectively over the same period [5]. Technological change has been shown to be the principal reason for these large differences in rates of economic growth. In most of these studies an expanded production function

$$Q = f(K, L, M, T)$$

(where K = capital, L = labor, M = material, and T = technology) has been used to determine the contribution of technological change to economic growth by computing the portion of economic growth which could not readily be explained otherwise, or by use of more traditional production functions.

Another issue which makes it difficult to determine the economic impact of technological change at the macro level is that many technological developments are new and not improvements or changes of existing technologies. In fact, there are many technologies now which serve a use that was not recognized as existing before. Therefore a comparative analysis of the economic impact of technology may be quite difficult, particularly if the impact is at least partially due to the introduction of new technologies which did not replace existing technologies but was the result of new discoveries which in turn generated new demand.

At the micro economic or individual firm's level the effects of technological change are usually more easily identified and directly tied to causal relationships. For example, product improvements or the introduction of new product technology can often be shown to directly generate a larger demand for a product, and possibly even larger market share, while changes in process technology and resulting productivity improvements will generally cause a reduction in costs and particularly marginal costs. If the marginal revenue and demand curves remain constant such a change in productivity resulting from technological change and not just learning would increase the output at which profit is maximized, making more of the output available at a lower unit price. Technological change at the micro level therefore often serves to implement macro or micro economic policies.

As shown in Section 7.4, the microeconomic analysis of technological change at the product or process (micro) level is complex and requires not only comparison of marginal costs but evaluation of the impact of technological change decisions on fixed costs per unit of output as well as residual fixed costs at the time of technological change. The opportunity cost of money and the method of depreciation of the investment in new technology at acquisition therefore plays a significant role in the determination of the change in production costs, and as a result the effect of the choice, timing, and rate of introduction of technological change.

Technology emerges from discovery and invention into innovation, a process which transforms the ideas discovered into useful, marketable, producible outputs. Fundamentally, innovation anchored in the basic results of discovery or invention is an inherently disorganized and often sketchily planned activity. The process of innovation requires various driving forces such as market demand, technological and financial capability, competitive position, and more - all of which vary constantly. As a result, innovation is usually a self-organizing stop and go process.

Contrary to popular belief, that technological development which includes R&D leading to discovery or invention and innovation, is a rational, predictable process which can be regulated and controlled. The reality is that technological development is erratic and seldom predictable. This, notwithstanding, attempts to manage it through planned research, development, and innovation.

Some of the most important technological discoveries were made by change and many important discoveries, once made, were not subjected to effective innovation by which they could be transformed into really useful applications. In fact, as noted by Sahal 9170, "no innovation of any significance has been made wholly by design". The same applies to a somewhat lesser degree to inventions, many of which are the result of chance discoveries. Important inventions or innovations often occur as "accidental by-products" of a mainline of research or innovation. Synthetic rubber, for example, was the result of research into the development of inexpensive anti-freeze for automobiles.

Technological developments and thereby technological progress is a cumulative process of adding to knowledge. It is, as noted, largely unplanned and disorganized. It involves chance and is subject to probabilistic laws. In fact, statistical analysis of past inventions shows 9170 that the probability of the number of discoveries per unit time is governed by a negative binomial distribution. Another important phenomena is that technical innovations do not occur singly but tend to cluster as innovation opportunities are triggered by one or more specific discoveries. In fact, sometimes actual avalanches are started by technological breakthroughs.

Economic conditions may impede such developments, particularly when they are most needed, during an economic downturn. Data indicates that most fruitful periods of technological advance occur during periods of economic recovery and recession and not during periods of economic prosperity and depression. In fact, peaks in technological innovation usually occur at the beginning of the recovery and recession periods.

Technological inventions and discoveries then lead more rapidly towards innovation during these periods under prevailing pressures to return the economy to stability and prosperity. On the other hand, during prosperity, people are usually too occupied with harvesting and enjoyment of the fruits of prosperity to devote themselves to the risky efforts of innovation.

11.1 INTERNATIONAL FLOW OF TECHNOLOGY

As often reported, most great powers rose to a position of global importance or even dominance through economic and technological achievements. As economic growth has been shown to be largely the result of technological advance, technology plays a most important role in the global position of a nation in terms of its economic and political influence, effective control, acceptance as a leader, and the quality of life of its citizens. Nations are increasingly concerned with their international standing, in both absolute and relative terms. This standing, most often defined in macro economic terms such as GDP or GNP, is a measure of relative external and internal changes which are increasingly defined as well as measured in technological terms.

Governments consider it one of their principal responsibilities to advance the political and economic status of their respective nations. They now increasingly assume policies that encourage economic and political growth through acquisition of technology. But technology is not randomly developed or its development evenly distributed or available, as some theoretical economic models appear to assume. Trends in the historical developments of technological advances show that they are usually heavily concentrated in certain parts of the world, notably the developed countries. In fact, the economic development of todays' industrial countries was largely technology driven.

Today most technology is commercialized with property rights attached to most economically attractive technological developments. As a result, transfer and acquisition of technology generally requires purchase and sale of right to technology. As technology development or acquisition is an essential precondition for economic development, and payment and expenses are required to acquire, develop, and transfer or introduce new technology, developing countries, who are usually in a weak bargaining position, often pay or spend more than others for such technology acquisition or transfer. The process of technology development or acquisition by developing countries is particularly expensive as technology changes continuously and developing countries often acquire and transfer technology when it is already near maturity and ultimate obsolescence. The method used in technology transfer influence the terms and costs of technology acquisition, which in turn are among the most critical issues in economic development.

11.2 COMPARATIVE MANAGEMENT OF TECHNOLOGICAL CHANGE

Firms considering introduction of new technology often operate in distinctly different environments, are subject to different constraints, and use different approaches to the management of technological change. External and internal constraints, and other impeding factors as well as complex procedures may also influence technological change decisions and resulting impacts. Therefore, the environment in which technological change is made should be evaluated to explain the large differences in the effectiveness of technological change experienced. This includes government involvement in and government aid towards technological change decisions. Similarly, levels of societal and cultural acceptance of technological change should be appraised or at least evaluated.

Procedures used in the determination of technological change requirements, discovery of new opportunities, and how objectives for technological change are set should similarly be studied. The approach to technological change decisions used by individual firms, including methods used to acquire, introduce, and use the new technology are similarly relevant. The major management parameters of concern to an industrial firm contemplating technological change are: ownership; company structure; management/labor relations; labor organization; work rules ; acceptance of technical and work content change; drive towards productivity improvement; benefit sharing; management decision procedure; employment security; government involvement; average time from identification of need for technological change to decision to invest; and, usual method of technology acquisition, as shown in Table 11.1.

Interviews and questionnaires can be used to determine how technological change decisions are made and who makes them. Similarly one can identify constraints which inhibit technological change decisions [Table 11.2]. Incentives and disincentives provided for such decisions should also be identified and evaluated. Finally, the methods used to implement decisions and the external and internal factors influencing technological change decision must be studied. During the interview typical questions posed would deal with:

1. How were technological needs or voids identified?
2. Who usually identified such needs?

3. How was an opportunity for technological change discovered?
4. How were opportunities evaluated?
5. How are, as a result, technological change decisions made?
6. Who made the technological change decisions?
7. What were the constraints on technological change decisions?
8. Where any incentives provided for technological change? What were the incentives and how were the incentives offered/provided?
9. What affects technological change decisions?
10. How are technological change decisions implemented?
11. What limitations on the rate of change, rate of introduction, and rate of switchover use existed and how were these limitations overcome?

Characteristic \ Shipyard	A	B	C
Ownership	Private, large holding company, publicly owned	Government with small minority share privately owned	Private, large conglomerate closely held
Company Structure	Subsidiary of corporation	Government enterprise	Corporation
Management-Labor Relations	Integral	Adversarial	Cooperative
Labor Organization	Company union	National union	Plant union
Work Rules	Negotiable	Rigid	Flexible
Acceptance of Technical and Work Content Change	Negotiated and reviewed by peer groups	Usually resisted by labor if perceived as labor saving	Universally
Drive Towards Productivity Improvement	Consistent drive towards improvement	Resistance	Relentless drive
Benefit Sharing	Year end bonus	None	Bonuses and other benefit sharing
Management Decision Procedure	Consensus - peer committees at various levels	Authoritarian	Explanatory with feedback and peer review
Employment Security	Permanent with exception of 30-40% of labor supplied by subcontract	Hourly except for supervisory and management personnel	All permanent
Government Involvement	Cooperative	Decisive - review all major decisions	Advisory and sometimes directive
Average Time from Identification of Need for Technological Change to Decision to Invest	Six months	One year	Three months
Usual Method of Technology Acquisition	Partially bundled equipment purchase	Purchase of components with bundling done in house	Bundled turn-key

Table 11.1 Major Management Parameters

The questionnaire (Table 11.2) is used to simplify the interview information.

The organizational structure of each firm should be analyzed and the management control systems studied in terms of decision sequences and the levels of management at which technological change decisions were made.

11.2.1 *Comparative Management Factors*. The management of technological change is affected by both exogenous and endogenous factors. Such exogenous factors include government control and regulations, dependence on government financing, uncontrolled market prices, input factor (labor and material) unit costs, exchange rates, and others. Many firms compete throughout an area or even the world, and often single firms have no significant share of the total market. Under such conditions management can neither affect market prices nor input factor costs, such as the unit costs of material or labor. Similarly, it cannot usually affect government policy or exchange rates, at least in the short run. While some endogenous factors, such as labor allocation, capital investment, utilization of process equipment, and training, are under the control of management, others such as work rules or organizational structure and more are usually outside its control, at least in the short run.

Some exogenous factors such as market demand and price affect firms in a particular market equally, others affect each firm differently. Similarly, while it is usually assumed that all firms have access to information on technology, the approaches to acquisition and introduction of technology available to them usually differ widely. Even access to technology information is becoming quite discriminatory, as both availability and ability to use or interpret information becomes more complex and expensive.

The differences in productivity or cost improvement of a firm can often be explained by difference in management approach to the use of information and endogenous and exogenous factors which influence or affect productivity or costs.

Qualitative and comparative analysis of the management of technological change is designed to indicate how technological change works in practice, and what endogenous and exogenous factors influence technological change decisions under different conditions. The objective is to improve the understanding of effective management of technological change at the product and process level. Management of technology effectiveness can be shown to be a function of:

1. organizational factors;
2. technological factors;
3. market factors;
4. environmental factors;
5. cultural factors; and,
6. political factors.

These factors are not mutually exclusive and, in fact, influence each other significantly. Cultural factors will usually affect political, environmental, and organizational developments which in turn affect management decisions affecting market and technological factors. Yet market factors, such as market share and price, also affect technological decisions. In other words, these factors are not only mutually interdependent, but also influenced by feedback effects. They can be expressed organizationally in terms of macro variables such as organizational structure and technology, and at the micro level by individual and group behavior of people such as by their attitudes, leadership, values, and more. Childs [16] concluded that macro level variables tend to become more and more similar across various cultures, while micro level variables are tending to maintain their cultural distinctiveness. Yet culture affects both macro and micro variables.

Technological Change and Equipment Purchase Decision Making Procedure	Individual or Group with Title	Time Since First Suggestion
1. Where do suggestions for capacity/technological changes originate?		
2. What happens to these suggestions? Are they discussed by a peer group or just presented to another level in the hierarchy?		
3. Who in your organization analyzes such suggestions and to whom are findings presented?		
4. Who else gets copies of the findings?		
5. Who in your organization translates the findings into recommendations?		
6. Are these recommendations specific enough to allow purchasing to ask for bids or quotations?		
7. Who makes the financial/cost analysis before submission to the decision maker?		
8. Is there a need to obtain outside (government) approval? Always or only if expected cost exceeds a certain amount of purchase from particular source?		
9. Who may impose constraints on technological/ capacity change decisions - workers, unions, managers, government, etc.?		
10. Who decides on the degree of bundling of new technology/capacity?		
11. Is there a need to involve labor/union/ government in decisions affecting changes in work rules, work content, etc.?		
12. Who makes the final decision on investments in new technology/capacity?		
13. Who, if anyone, approves or ratifies the decision - peer group, board, government, finance committee, union, etc.?		
14. How long does tendering approval, acceptance, contracting, etc. take?		

Please attach an organizational chart and indicate the functions of each box on the decision making process. Also please estimate the time required for decision at each level.

Table 11.2. Comparative Management Questionnaire.

Cultural differences and values more than differences in organizational variables appear to affect differences in micro variables and resulting differences in management approaches and leadership styles. As an example, while some cultures focus on interpersonal relationships as a principal determinant of their management approach, others stress group and cooperative effort to a much larger extent, and others yet foster class consciousness and egocentricity which, in turn, results in strict hierarchial management.

Cultural backgrounds in general affect the degree of individuality, respect for authority, desire for conformity, and motivation of workers and managers. While some cultures encourage cooperation, others emphasize competitiveness. Worker expectations and the meaning of work to the workers are largely influenced by cultural, social, and structural variables.

Management must be sensitive to such worker perceptions and show its concern through its leadership approach, if it is to be successful and particularly if effective technological change is to be achieved.

Cultural variables such as value and attitude are determinants of management behavior, and are usually considered moderator variables. Macro variables such as market (price) and technological variables are often similar across different cultures. Others, such as environmental variables, will influence the moderator micro variables which ultimately influence management decisionmaking and thereby management effectiveness. Macro variables therefore influence management through moderator micro variables. In the analysis of comparative management we must ensure that the experimental results can be attributed to the operationalization of culture as one of the independent macro variables, and search for both similarity and differences in comparing organizations and cultural variables to determine how management and worker styles differ across cultures.

While there are some inconsistencies in the results of Hofstede and Bass and Burger, it was possible to compile from them a general thesis of management values, attitudes, and practices for Japan and Spain, for which adequate findings were reported. Data for Korea were obtained from several working papers on Management Effectiveness published by the World Bank, and from the works of England et al [14] who found a high degree of pragmatism among managers in all the countries under consideration here. (Pragmatic values emphasized productivity, profitability, and achievement and correlated positively with career success.)

Most of the comparative management studies reviewed were based on a questionnaire approach. Such experiments often suffer from problems of interpretation by, and the cultural incisiveness of, the respondents. Furthermore, there are major differences in classification and specification of both macro and, more importantly, micro variables. Among investigators, it is therefore only possible to establish a consistent thesis in general terms. While this sometimes provides a useful test of the findings of comparative management research, it makes a specific comparative analysis of management behavior and practices affecting technological change effectiveness rather difficult.

11.2.2 *Rationale for Comparative Management Analysis*. The findings of comparative management research should be tested against ex-ante presumptions derived from studies of comparative management performed by other investigators.

Some researchers such as Ronen [6] have attempted to cluster countries according to similarities in cultural dimensions such as work goals, values, needs, job attitude, and management objectives. The discriminant validity of these variables is supported by the fact that cluster members as a group discriminate on the basis of various locational and cultural issues. While Latin, Nordic, Arab, Germanic, and similar clusters can be well defined, many newly industrialized countries such as Korea and Israel do not follow the general rule and are therefore usually grouped as independents. It is interesting to note though that even Japan, which has been firmly established as an industrial country for a long time, does not fit into the cluster of other mature industrialized countries such as the USA and the Western European nations. The Negandhi and Prasad [7] model does incorporate cultural variables, but subordinates their effects to those of management philosophy. This approach, while meaningful in the comparative management study of multinational firms, is also not appropriate for purposes of comparison of management of technological change.

Cross-cultural studies of management behavior and leadership began with Haire et al [8] who presented the first systematic study in 1966, and were followed by the development of Fiedlers [9] contingency theory, Broom and Yettons [10] normative model, and House and Mitchells [11] path-goal theory. While these theoretical studies provide an important analytical framework, they do not address the particular issues of comparative management pertinent to this research.

Comparisons of management behavior, style, and practice, as well as comparative management analysis under different cultural environments, by Bass and Burger [12] and Hofstede [13], on the other hand, provide meaningful presumptions. Their findings show that values are major cultural variables which affect management behavior and practice. They similarly showed that there is no culture-free context of organization. It was also found that cultural factors affect methods and effectiveness of communications and interpersonal relations which differ significantly among nations and cultural groups.

England et al [14] on the other hand found a high degree of pragmatism among managers in all the countries under consideration here. (Pragmatic values emphasized productivity, profitability, and achievement and correlated positively with career success.)

Most of the comparative management studies are based on a questionnaire approach. Such experiments often suffer from problems of interpretation by, and the cultural incisiveness of, the respondents. Furthermore, there are major differences in classification and specification of both macro and, more importantly, micro variables. Among investigators, it was therefore only possible to establish a consistent thesis in general terms. While this provides a useful test, it makes a specific comparative analysis of management behavior and practices affecting technological change effectiveness rather difficult.

11.2.3 *Criteria for Measurement of Management Processes*. The setting of norms for measurement of management effectiveness has occupied many researchers. For example, Negandhi [15] proposed to use indices for management processes, personnel practices, and decentralization in the study of management philosophy. Similarly different factors have been proposed as measures of management effectiveness. As the use of financial data as sole or even dominant measures of management effectiveness is unsuitable, numerous lists of behavioral, organizational, and other factors have been proposed as management effectiveness criteria. Most of these are based upon the availability of data from industrial firms studied. Some of the criteria proposed though are controversial or too narrow to be of use in comparative management analysis.

Behavioral criteria such as worker morale and satisfaction may sometimes lead to controversial results when used as measures of worker attitude and productivity or output. Worker morale and satisfaction are really dependent variables, closely related to interpersonal relationships which may often be interpreted as a causal variable.

In some environments and at some levels the need for security is more important than in others which permit larger utilization of advanced, usually labor saving, technology without worker alienation.

The need for prestige, self realization and esteem also affects management's approach to the method of acquisition of new technology. The high level of need for prestige and higher levels of need for esteem in some environments can be expected to make it harder for management to import bundled technology.

11.2.4 *Technological Change and the Management Organization*. Technological decisions are made at various levels in an organization, from top corporate managers to product or process section managers. As a result, such decisions are often made at different levels of an organization which may actually lack the experience in and knowledge of the technology of concern and its expected uses or applications. The structure of an organization, its culture, internal communication links, background of decision makers, accountability, and usual planning, budgeting, and resulting decision horizons all affect the management of technological change.

Highly technical firms organized in matrix structure or with a small number of management levels provide for easy accessibility of top decision makers by technological experts as well as potential developers or users of the technology are usually able to make much more rapid, timely, and effective technological change decisions, than more hierarchical, short term planning, and highly structured management organizations.

The most successful, high technology organizations in fields where process and product technology changes vary rapidly are loosely structured, have a very flat organizational chart, with few levels of management, devote a lot of attention to communication, involve all levels of the organization in decision making, have an open door policy, plan principally for the long term and develop short to medium plans not to attain a short-term objective, but to support the long-term plan and its objectives, are managed by technical staff with all senior positions filled by experts in the process/product field of the company.

11.3 COMPARATIVE MANAGEMENT MODELS

Japan

The Japanese government exerts indirect pressure on industry in general by providing investment loan support, export financing, and funding of research and development through a number of government departments and joint government/industry agencies. It also assists in setting national industrial policy in terms of investment (disinvestment) and, as a result, output. These policies developed jointly by industry and government influence technological development (R&D) and market share of Japan (and even individual Japanese industries) directly. They affect management of technological change at the process level only indirectly.

Cultural traditions in Japan emphasize close cooperation not only among units of large industrial and service companies, but also among industrial units of different companies in the same field. As a result, industrial firms usually share experience and other information. Data on technological developments is therefore quickly diffused, and industries are able to rapidly respond to technological change opportunities or needs.

There is also general support of technological change by the community which perceives the integral role of enterprises in its own development. As a result, there are generally few exogenous obstructions to technological change decisions, which are usually assumed to further the general well being and progress.

The institutional structure of most Japanese firms is hierarchical as are its decision making procedures, though peer or group decisions usually prevail. Each level of the organization has well-defined decision making powers and delegation of responsibilities. Job security, 'lifetime' employment, and profit sharing incentives usually assure not only peer participation and input into the decision making process, but also full cooperation in the planning and implementation of technological change, but at all levels of the organization.

Need for technological change is identified by production or work center management, who are also responsible for cost control. The recommendation for change is made on the basis of need for more (or different) capacity as well as improved productivity (or cost) at the process level.

The management of technological change starts with such a recommendation which usually identify

1. the process, product, or service to be changed,
2. the new capacity required or demand to be satisfied,
3. the choice of new technology, or
4. planned use of the new technology.

These recommendations derived through consensus by peers at the work center and production management levels are then submitted to department management which reviews them with other department managers. When a consensus is achieved at this level, the recommendations are usually transferred to management if they involve substantial investment, where they are evaluated in terms of consistence with policy.

Management will usually decide on issues such as size of investment, timing of order, bundling rate, and lead time, while department and work center management makes the decisions affecting training, utilization of new technology, and choice of technology.

Management decisions (timing, bundling rate, and so forth) depend to a large degree on policy objectives, both company and industry, as well as aggregate exogenous factors such as world demand and price, while department or work center management decisions are based on capacity need and work center productivity or cost.

The choice of technology is usually based on readily available status and performance information. Training of workers is nearly always in the requirements or use of the new technology. As a result, utilization of new process technology is gradual, as the number of people who can be trained at the same time is often quite small.

This model of management of technological change offers good control of timing of technological change and effective technological choice of decisions, but delays the effective utilization of the new capacity. This is verified by the results of the qualitative analysis which showed that Japanese industry often have the earliest timing of technological change, a high bundling rate, reasonable lead time, but only a gradual training of workers and put into use of the new technology. This model performs reasonably well but results in a smaller than possible change in the experience curve between old and new technology.

Spain

Most major Spanish firms are managed from a central headquarters with an operating/production management located at the site. Local management is responsible for day-to-day operations, with all planning, engineering, and investment decisions made at the central headquarters. As a result, there is little delegation of authority.

Central management reports directly to the board which approves budget, sales, and operating plans. Industry is in fact run like a government agency with headquarters. The objectives of headquarter management are to operate strictly within guidelines.

Its principal goal is to further the macro economic, industrial, and trade policy of a company. As a result, preference in procurement is given to sister enterprises independent of quality or costs. Also company enterprise buyers are given certain advantages. Most senior headquarter management personnel are on civil service status. Workers at the plant are highly organized by a national union which imposes strict working rules and other conditions. The company has no incentive system in place.

Technological change decisions are made by the central management based only in part on a 'post factum' evaluation of plant work center performance. In most cases technological change proposals are made by senior management such as the director of engineering, planning, or sales based not on an actual need identified at the yard but based on their findings of a desirable development in terms of advanced technology, increased capacity, or improved productivity. The proposal is then studied by the professional staff of the planning and engineering department with some inputs by the sales department. The results of that study are presented to the management consisting of department directors who then decide if to go ahead with the investment.

Production management and personnel is usually informed of such technological change after the fact. The reason given for this approach is that involvement of plant personnel in the process would result in blocking of any plan or proposal for change because of strong union presence and control. Preference for other enterprise products causes a low bundling rate for purchases of equipment made elsewhere, as the contribution of enterprises in any purchase is maximized. Procurement is, as a result, often limited to proprietary components with structural, foundation, and other basic parts built or supplied by a sister enterprise or the plant itself. This causes major delays in the lead time and also sometimes affects the quality of the installed equipment.

Training of workers is very gradual and on the job. It can usually only be done with lengthy negotiations with the union and various workers' committees.

This model of top-down management of technological change often causes late timing of decision to add capacity or new technology, or timing which is not properly coordinated with real need. Similarly process lead time is extremely long and the bundling rate rather low. Decisions on technological change are greatly affected by government policy such as foreign trade and economic development, as well as domestic employment and local development policy. Local management which only controls training of workers and equipment utilization is even constrained in these decisions by labor contracts negotiated by headquarters. As a result, training and utilization is often delayed and introduced at an extremely slow pace.

This type of model of management of technological change is quite inefficient as it does not allow effective response to changes in demand, cost, and availability of new technology.

Korea

Most large plants are privately owned and part of larger conglomerates or industrial companies. While government influences major industrial decisions by setting economic and financial policy, the company is largely free of government control, interference, and support. There is a large degree of delegation of responsibility with each level of management authorized to make decisions appropriate to its responsibility.

Proposals for technological (or capacity) change decisions originate at the shop or work center level where the need, choice of technology, and scale or level of investment required is analyzed and developed into a recommendation which is then presented to the department (usually production department) head to review among his peers.

Decisions of medium-sized investments can usually be made at this level. If the investment and proposed change is large or affects output and/or labor (work center) manning or organization, then the recommendations are presented to management for final decision.

As department heads meet with work center managers as well as management daily, this process takes very little time. The work center and its production department are responsible to not only make the choice of the technology but also analyze its expected impact, develop a detailed procurement and implementation plan, as well as a plan for the training of workers and new technology utilization.

Few environmental factors interfere with this management of technological change. Government is basically not involved and the community-at-large is generally cooperative. With workers permanently employed and provided with both very extensive benefits and monetary incentives, full cooperation at all levels of the work force and management is generally achieved.

This model allows for effective timing of technological change decisions, a very short lead time, well-planned worker training programs, and large-scale utilization of new technology from the start.

The bundling rate is similarly usually very high in recognition of the fact that any money saved by lower levels of bundling will be more than balanced by losses resulting from delays in installation and effective utilization of the new equipment.

Korean firms often use bundling to assure effective knowledge and technology transfer and to allow training of its staff and workers by the technology supplier. Finally, bundling shifts the responsibility for the effective operation of the equipment to the supplier.

11.3.1 *Model Evaluation*. The three models of management of technological change differ significantly in the levels at which management decisions are made, the timeliness of the decisions, and the approach to decision making. To be effective, the management of technological change must assure that decisions are made at a level of the management structure which has the best knowledge of and experience with the issues involved. It is therefore not surprising to note that the Korean model - among all the models - exhibits the shortest lead time, highest utilization, largest bundling, and largest number of workers trained prior to installation of new equipment or introduction of new technology.

Quantitative analysis [Ref. Chapter 5] has shown that growth of utilization of new technology is the most important factor influencing improvement in the learning curve. The degree of utilization of new technology, including the rate of actual putting into use of new capacity, is largely a factor of the participation of the work center staff in the decision making process, including their involvement in the use planning of the new equipment. Similarly the bundling rate, which is a dominating factor in the determination of the timing of the slope change, is higher in a model where the decisions are made at the production and not company management level.

11.4 IMPACT OF TECHNOLOGICAL CHANGE ON WORKER AND MANAGER

Technological change has become not only more frequent for most manufacturing or service organizations, but has become a near continuous process. Technological change today always involves change and improvement in information handling and use. It also requires continual learning at all levels of an organization, from worker to top manager. Technological change also affects organizational structure and hierarchy and nearly always leads to simpler structures and a reduction in the number of hierarchical levels. The use of advanced information technology often leads to better management and more effective decisionmaking at all levels of an organization. It also assures more effective control and integration of new process and product technology.

It has generally been assumed that worker resistance to the introduction of new process (particularly automated process technology) and information technology was largely based on fear of the loss of jobs, and the higher skill requirements demanded by the new technology. It has more recently been recognized that management, and particular middle and upper middle management, resists such change even more forcefully. They often fear that automation and advanced information technology will allow lower level managers and even workers to make decisions which were traditionally theirs. New technology can often turn workers into managers and allow low level managers to make high level decisions. The clash between new technological requirements and old management attitudes and procedures is now becoming more acute, as managers who for long explained the advantages of technological change in terms of manpower reduction and productivity, now often find that the jobs eliminated may include their own, as flattened hierarchies, improved information flow, more decisionmaking by expert systems programs, and more, make many traditional management positions or functions redundant. Computers can now make many decisions more effectively by use of large flows of information which they can use more efficiently, timely, and accurately at both the worker and manager decisionmaking levels, which in turn eliminate the need for manager or worker decisions. The trend of replacing workers as well as managers by technology will continue, although it will not, as is often assumed, lead to large-scale reduction in job opportunities. New jobs will be generated which will be different and largely involved in the development of new technological hardware and software.

11.4.1 *Labor/Management Cooperation.* Technological innovation by a firm requires labor/management cooperation to be successful. In recent years many U.S. unions have had to face the issue of traditional confrontation versus new cooperation with management for mutual benefit, including job security and improved competitiveness. This issue now pervades the politics and policies of the United Automobile Workers (UAW), as an example, which since early 1980s, had suspended its historic mission of pressing for ever better economic benefits for its members by confrontation, so as to concentrate on job maintenance and improvement through cooperation. This meant agreeing to cooperation between hourly workers and management, including acceptance of many innovations, different work rules, job contents, and most importantly acceptance of the team approach where everyone did the level best for the good of the firm. The UAW is now facing a political and ideological challenge from dissidents opposed to the new union strategy which has safeguarded members' job security by helping improve the firm's or industry's competitiveness against foreign automakers. The new approach has given the union a say in policy and strategy, including new approaches to health, safety, and other social benefits and joint production scheduling, job assignments, and total quality management.

Innovations were introduced more effectively and in a more timely manner in product design, process improvements, and organization of manufacturing. Labor management cooperation assured not only better acceptance but also more effective use of innovative technology. This approach, new in U.S. industry, has resulted in workplace restructuring, a new spirit of togetherness or jointness, and a recognition of mutual benefits and common objectives by labor and management. Structural adjustment in U.S. industry has not come easy or without cost. Older workers are often replaced or leave as they find it difficult to adjust to the new environment which attracts younger workers without a burden of historic mistrust.

11.5 GOVERNMENT ROLE

An important consideration is the effect or role of government in the management of technological change. Governments may assume a non-obtrusive supportive role, a participatory enhancing role, or a controlling role. These different approaches affect the timing, scope, decisiveness, and approach to investments in technological change. The unimpeded competitive environment in which some firms operate allows technological change decisions to be made mainly on the basis of commercial and operational factors. The management of other firms, while free to make its own decisions, coordinate much of their technological change investments with government and industry bodies. In other words, government's role may be non-participatory when it only provides support in facilitating acquisition of new technology and is informed of technological changes only after the decisions and commitments were made. In other countries, on the other hand, government bodies work closely with industry and industry associations in the development of technological strategies. Finally, there are countries where governments control major technological change decisions made by industry and in fact superimpose themselves directly or indirectly in many technological change decisions.

11.5.1 *Developing Country Technology Gap*. Developing country governments are in a bind when it comes to technological development. They have to advance technologically in the long run to catch up with developed countries' standard of living and per capita or gross domestic income. Yet they chronically lack employment, with unemployment rates often higher than 50%. Introducing new process technology nearly always implies a loss of employment and an increase in the level of skill required to perform a job. But skilled or trained labor is extremely scarce in developing countries. Altogether developing countries make up over 83% of the world's population, less than 13% of the world scientists and engineers are in the Third World, and they are concentrated in only a few countries in East Asia, Brazil, India, and Mexico.

Also new technology requires large amounts of capital which most developing countries lack. They also lack effective access to capital markets, and are usually unable to attract investment or capital at competitive terms because of:

1. Risk to investment due to commercial, political, and regulatory factors, including government interference, takeover, etc.
2. Lack of effective management
3. Lack of adequate skilled labor and imposed labor hiring requirements.
4. Insufficient marketing and market controls
5. Restrictive investment and foreign exchange laws, as well as large foreign debts which may cause restrictions on repatriation or repayment of capital
6. Inadequate local banking and financial institutions.

Developing countries therefore face difficult choices. If new laborsaving technology is introduced, then even fewer jobs (though higher skilled and better paying jobs) will be available, increasing the already dangerously high level of unemployment. If they do not introduce new technology, their products will not be competitive in international markets, because of price as well as quality and product form. The shift in world consumption and technological innovation have cut demand by industrial countries for traditional raw materials. Over 50% of all developing countries though continue to depend on raw material exports for the major part of their export earnings.

Furthermore, new product and process technology is already making developing countries less attractive for investment or as sources of manufactured goods as the low labor content, high skill and capital requirements make it increasingly more attractive to maintain advanced manufacturing plants in developing and newly industrial countries (NICS).

Technological advances therefore today increase the barriers transforming developing countries into industrialized countries, a fact which may have long lasting effects on the world, by permanently disadvantaging a large segment of humanity and increasing the differences in living standards and living styles even further.

At the same time the world is moving towards more interdependence in trade and the development of large economic blocks in Asia, Europe, and North America. As trade among industrial countries (including NICS) continues to grow, trade among developing countries is practically non-existent.

Trade between developing and developed countries has grown at a lower rate than world trade as a whole in recent years. While this is in part due to low commodity prices which reduce both the value of developing country export earnings as well as their ability to pay for imports, the volume of this trade has similarly stagnated because of import substitution and other technology-induced factors. Yet the economic performance of developing countries will influence the ability of developed countries to take advantage of technological advance and continue their economic growth. Unless markets expand in the developing countries which account for over 80% of the world's population, developed country trade will ultimately stagnate as well. Even more assertive may be the political pressure for greater equity and better wealth redistribution. Unless something is done to at least move in this direction, then the world debt crisis of 1988 will soon pale in comparison with debt and trade imbalances at the end of this century.

The only real opportunity lies in advancing developing country technological prowess. If these countries are not to be left further behind in economic and human terms, their national scientific and technological capability must be improved.

An increasing number of their products are no longer suitable for or acceptable to industrial country markets. They must develop suitable technologies which offer some competitive advantage using particular local skills, materials, or proximity to markets. They will have to learn how to deflect negative impacts of technological change through a changeover to use of greater skill and a resulting reduction in income disparities. The effects of better educational, training, communication, health, and other social programs and related investments or resource use will have to be explained and acceptance developed.

Technological change may have undesirable short-term effects on developing countries, but in the longer run it will help in eliminating the endemic poverty, large income differences, and unacceptable levels of unemployment which are now the mark of most of the developing world. This will bring in its path greater stability, improved human well being, and better understanding among man. As a result, the world may be a more peaceful and happy place to inhabit by future generations who use earth resources wiser and relate to each other more effectively than we do today.

References

1. Solow, R. M., "Technical Change and Aggregate Production Function", The Review of Economics and Statistics, Vol. XXXIV, August 1957.

2. Hicks, J. R., "The Theory of Wages", MacMillan and Sons, London, 1963.

3. Harrod, R. F., "Economic Dynamics", MacMillan and Sons, London, 1963.

4. Jones, H., "An Introduction to Modern Theories of Economic Growth", Thomas Nelson and Sons, Ltd., London, 1975.

5. Council of Economic Advisers, "Economic Report to the President, 1984". U.S. Government Printing Office, Washington, DC, 1985.

6. Ronen, S., "Comparative and Multinational Management", Wiley, New York, 1986.

7. Negandi. A. L. and Prasad, S. B., "Comparative Management", Appleton Century Crafts, New York, 1971.

8. Haire, M., Giselli, E. F., and Porter, W., "Managerial Thinking - An International Study", Wiley, New York, 1966.

9. Fiedler, F. E., "A Theory of Leadership Effectiveness", McGraw Hill, New York, 1967.

10. Vroom, V. H. and Yettons, P. W., "Leadership and Decision Making", University of Pittsburgh Press, Pittsburgh, 1973.

11. House, R. J. and Mitchell, T. R., "Path Goal Theory of Leadership", Journal of Contemporary Business, 1974.

12. Bass, B. and Burger, P. C., "Assessment of Managers: An International Comparison", Free Trade Press, New York, 1979.

13. Hofstede, G., "Measuring Hierarchical Power Distance in 37 Countries", Working Paper 71-32, European Institute for Advanced Studies in Management, 1981.

14. England, G. W. and Lee, R., "Organizational Goals and Expected Behavior Among American, Japanese, and Korean Managers - A Comparative Study", Academy of Management Journal, December 1971.

15. Negandhi, A. R. and Prasad, S. B., "Comparative Management", Appleton-Century Crafts, New York, 1971.

16. Childs, J. and Kieser, A., "Contrasts in British and West German Management Practices: Are Recipes for Success Culture Bound?", Conference on Cross Cultural Studies on Organizational Functioning, Hawaii, 1977.

17. Sahal, D., "Invention, Innovation, and Economic Evolution", Technological Forecasting and Social Change, Vol. 23, 1983.

Bibliography - The Economics of Technological Change and Comparative Management

1. Dewar, A. and Simet, D., "The Quality Circle Guide to Participation Management", Asia Productivity Center, 1982.

2. Cummings, P. W., "Open Management, AMACOM, New York, 1980.

3. Blau, A. V., "Approaches to the Study of Social Structure", Free Press, 1975.

4. Hayes, H. R. and Clark, K. B., "Why Some Factories are more Productive than Others", Harvard Business Review, September/October 1986.

5. Packer, M. B., "Measuring the Intangible in Productivity", Technology Review, February-March 1983.

6. Harbison, F. and Myers, C., "Management in the Industrial World: An International Study", McGraw Hill, New York, 1959.

7. Farmer, R. N. and Richman, B. M., "Comparative Management and Economic Progress", Homewood, IL, Irwin, 1965.

8. Negandhi, A. R. and Prasad, S. B., "Comparative Management", Appleton-Century Crafts, New York, 1971.

9. Chowdry, R. and Pal, A. K., "Production Planning and Organizational Morale", in A. H. Robinson et al (Eds), "Some Theories of Organization" Dorsey Press, Homewood, IL, 1960.

10. England, G. W. and Lee, R., "Organizational Goals and Expected Behavior Among American, Japanese, and Korean Managers - A Comparative Study", Academy of Management Journal, December 1971.

11. Haire, M., Ghiselli, E. F., and Porter, W., "Managerial Thinking: An International Study", Wiley, New York, 1966.

12. Bass, B. and Burger, P. C., "Assessment of Managers: An International Comparison", Free Trade Press, New York, 1979.

13. Hofstede, G., "Measuring Hierarchical Power Distance in 37 Countries", Working Paper 71-32, European Institute for Advanced Studies in Management, 1981.

14. Hofstede, G., "Motivation, Leadership and Organization: Do American Theories Apply Abroad", Organizational Dynamics, Vol. 9, 1983.

15. Childs, J. and Kieser, A., "Contrasts in British and West German Management Practices: Are Recipes for Success Culture Bound?", Conference on Cross Cultural Studies on Organizational Functioning, Hawaii, 1977.

16. Ronen, S., "Comparative and Multinational Management", Wiley, New York, 1986.

17. Fiedler, F. E., "A Theory of Leadership Effectiveness", McGraw Hill, New York, 1967.

18. Vroom, V. H. and Yettons, P. W., "Leadership and Decision Making", University of Pittsburgh Press, Pittsburgh, 1973.

19. House, R. J. and Mitchell, T. R., Path Goal Theory of Leadership", Journal of Contemporary Business, 1974.

20. Kim, L., Aaron, H., and Westphal, L. E., "Management in Korea's Expanding Industries", Development Research Department, World Bank Report No. DRD 61, Washington, December 1981.

21. Westphal, L. E., Kim, L., and Dahlman, C. J., "Korea's Acquisition of Technological Capability", Development Research Department, World Bank Report No. DRD77, Washington, April 1984.

22. England, G. W., Dhingra, O. P., and Agarwal, N. C., "The Manager and the Man", Kent Ohio: Kent State University Press, 1974.

APPENDIX 11A - SUMMARY OF COMPARATIVE MANAGEMENT OF TECHNOLOGY STUDIES

Similar firms - in size, output, and product - were studies in Korea, Japan, and Spain. The results of these investigations by on site observation and evaluation are summarized here.

Japanese Management

The Japanese firm was owned by a large, publicly-owned holding and trading company, and run as a subsidiary company. Management was mainly promoted from inside the firm or the company. The company was paternalistic and generally provided lifetime employment to its workers and staff. To manage fluctuations in workload, the number of workers was usually kept at 60-70% of requirements, with the rest of the work subcontracted. Subcontractors usually supplied workers only for certain functions within the firm. Only in isolated cases would a subcontractor perform the required work at his own firm. Permanent employment declined from 12,200 in 1972 to just over 8,320 in 1982. Labor was organized in a company union and labor shares in the profits of the firm. Other incentives were provided in the form of individual and group benefits.

Management decisions at all levels of management were largely made on the basis of consensus. Peer review planed an important role in all decisionmaking, including decisions on technological change. The government was cooperative and supportive and did not regulate or interfere with management decisions unless than required use of government resources. Investment and technological change proposals usually originated at the work center level. After achieving a consensus at that level using a peer review procedure, the proposal was submitted to the next level of management where consensus was sought again. This process was continued until it reached the level authorized to make the relevant decision or commitment. It usually took 3-6 months for a proposal for a change in technology to be acted on.

Once a decision was made, all levels of the organization were informed and credit was given to the originating peer group which then was usually involved in and given the responsibility for developing the implementation plan. As a result, workers most directly affected by a technological change formed an integral part of the team responsible for planning the details of the technological change, including technology choice, timing, scale, and more. Acceptance was therefore a foregone conclusion. New technology was generally bought partially bundled with the complete process technology supplied by the maker, but assembly and test were performed by the firm under the supplier's guidance. Training was usually performed by the firm itself and provided by its own training department on newly installed process equipment instead of at the equipment manufacturer's facility. Therefore, contrary to the practice in countries where a significant number of workers were trained in the use of the new equipment before its installation, the Japanese firm trained workers only after the introduction of the new equipment. This resulted in a more gradual run-in process for new technology, but may have provided advantages in training effectiveness resulting from the familiarity of the workers with the environment and the realistic use of the new equipment.

Company management provided general guidance with respect to sources of supply and methods of acquisition of new equipment. Similarly, various Japanese government agencies engaged in and/or funded technology research. There were a number of industry-government associations engaged in the exchange of information, joint research, and technology policy determination. Most Japanese firms in that industry appeared to share technological developments, research as well as experience results, and assisted each other through these associations in resolving technological problems.

Korean Management

This firm had a hierarchial organizational structure with department directors reporting to the Chief Operating Officer. They are responsible for major functional areas such as contract administration, production engineering, production planning, production, cost control, purchasing and inventory control, personnel, financial and accounting, marketing and public relations, and legal and government relations, each headed by a vice president.

Each department is divided into a number of major divisions which are, in turn, subdivided into sections and subsections.

Employment at the firm was 9,200 at the end of 1984, of which 7,320 were production workers, 880 clerical and security staff, 220 engineers, 400 designers, draftsmen, and programmers, and 380 management personnel, including 182 foremen and shift supervisors, 92 section and subsection heads, and 84 division and department heads. The remaining twenty-two members consisted of professionals, such as lawyers, marketing experts, etc.

One of the most striking observations was the positive attitude towards technological change and its effects. People at all levels of the workforce and staff continuously evaluated their performance and the effectiveness of the technology used and tried to improve on existing process technology or encourage change. Awareness of developments in, and the availability of, more advanced or better technology was fostered by discussions with foreign experts, training abroad, visits to foreign firms and, on occasion, the hiring of consultants. Most importantly, management established an environment which sensitized workers and staff to technological considerations and stimulated applications innovation to achieve better performance. In Korea, with little government-sponsored research in process development, emphasis was given to the maximization of performance in the use of imported process technology and the search for better foreign technology. This was quite different from the Japanese approach which was to initially copy foreign technology and then to develop local improvements in manufacturing process technology based on some, but not all, of the principles of the foreign equivalent.

Although the government did not encourage significant process R&D, it provided important incentives to invest in new technology. Similarly top management motivated workers and staff towards performance improvements by providing monetary, career, social mobility, and other incentives. As a result, both public and worker/staff attitudes were aimed at a continuous drive towards performance improvement and, when advantageous, technological change.

Determination of the decisionmaking process leading to technological change was difficult because many individuals at various levels of management and government and other factors contributed to such decisions.

Two distinct decision sequences could be identified. The internal decision sequence has its root on the shop level. Shop foremen met daily with their workers for a brief discussion of work tasks and performance during which peer discussion identifies issues of performance and quality. Problems are reviewed and foremen transferred information of experience gained by other work teams. Suggestions made and experience gained was brought to the attention of other foremen and the shop or work center manager who, in turn, would tell the foreman of experience in other work centers employing the same or similar processes. Workers or foremen who brought up interesting ideas for performance improvement would usually be asked to join the daily (afternoon) meeting of shop managers, in which senior management participated selectively.

Ideas for improvements are brought before these meetings by shop/work center foremen and managers who identify a need and often suggest either improvement in the currently used process or introduction of a new process. Information on new process technology is usually obtained from visits abroad, owner's representatives, and other sources. The implications of a technological change are discussed at this level. About half the technological changes were introduced for consideration by such a bottom-up process. The other half were proposed by management, often on the basis of perceived need and opportunity for performance improvements.

The decision to consider or reject proposals for technological change was made at the shop manager's level. If the proposal was accepted for consideration, one or more managers would be charged with the responsibility for investigating the impact, reliability, performance, and cost of such new technology, including identification and evaluation of alternative or competing technologies.

Their conclusions were reviewed by the group of shop/work center managers and then passed on to top management which, if the change, impact, or cost was significant, would confer with the appropriate government officials for concurrence.

A responsible top manager, usually the executive vice president in charge of production, would make the final decision.

The responsible shop manager would next be charged with the development of an implementation plan which included the establishment of:

1. acquisition schedule and budget;
2. layout and civil work requirements;
3. work and process flow changes;
4. work and job assignments;
5. training plans and schedules;
6. plans for installation of new process;
7. start up and test plans;
8. operational procedures and plans;
9. rate of introduction and replacement of other processes;
10. quality control and calibration; and
11. performance/productivity measurement.

The plan was reviewed by top management after discussion at shop manager's meetings and, if accepted, given to the procurement/purchase department for action. If the equipment to be acquired was costly and not specifically known, or if alternative technologies existed, a mission of managers and foremen would be sent abroad to investigate and recommend a choice which, if approved, would then be acted upon. The appropriate government officials were usually kept informed of the process if the change was significant or expensive.

Proposals or suggestions for technological change may also be advanced by government officials, outside consultants, owner's representatives, or others. Such suggestions will usually be subjected to the same procedure as suggestions advanced from the shop floor.

Management personnel at or above the rank of division head which included all shop managers, usually met twice a day, once before and once after the end of the day shift, to discuss current problems and outstanding issues.

In addition, small groups of concerned managers were assembled several times per week to discuss planning issues. Weekly summary division (or shop) reports and daily summary section reports were generated by a computerized management information and control system, and were reviewed at these management meetings. Adjustments to resource allocation were made at these meetings by balancing resource demand and supply on a shop-by-shop or division-by-division basis. Manpower, facilities, areas, and equipment was reassigned based on agreed upon priorities or needs in line with overall product completion or delivery requirements.

The average period required from the time a need for additional capacity and new technology was identified (usually at the work center level) to the final decision to invest in new technology was three months.

New technology was generally bought bundled on turn-key terms, which included training of personnel, installation, test, and supervision by the supplier during the run-in period. When ready, the new technology was used to the maximum reasonable extent consistent with work load. When new technology capacity was in excess of total output requirements, then all new production was shifted to the new capacity to take advantage of its higher productivity and, often, its larger experience curve slope. There were no limitations on the rate of introduction or switchover of new process technology.

Spanish Management

The government, through the Central Bank, was a majority owner of this firm and appointed most senior management located in Madrid, about 400 miles from the plant. Local management was responsible for day-by-day operations but had limited strategic, planning, investment, pricing or marketing decisionmaking power. Investment, labor relations, marketing, pricing, contract administration, and similar decisions were all made at headquarters. Also, engineering and design were handled at headquarters for all the plants of the company. The functional departments at the plant were headed by operating managers. There was very little delegation of authority from headquarters to plant management, and only routine questions and day-to-day operating management decisions were resolved at the local level.

Employment at the plant dropped from 16,860 in 1972 to 9,080 in 1982, as a result of a decline of orders. The plant was highly unionized with rigid work rules, and an adversarial relationship between union/workers and management.

The relation between plant and headquarter management was remote, strained, and bureaucratic. Major investment, pricing and marketing, as well as labor-related policies and resulting decisions, had to be reviewed and approved by government or the Central Bank. This included investment decisions affecting process technology or capacity change. The time required for action on a request by plant management was 6-12 months.

The plant itself was run very much like a government organization, with a strongly entrenched bureaucracy. As a result, there was very little drive towards productivity improvement or acceptance of technological change. About 80% of plant labor was permanently employed and organized by skill categories, with the rest engaged as casual labor.

There was very little, if any, feedback between workers and operational plant management on one side and plant management and headquarters on the other.

APPENDIX 11B - JAPANESE AND U.S. TECHNOLOGY DEVELOPMENT

The economic growth of and rate of technological development of Japan during the last three decades is generally associated with Japan's superior ability to recognize commercial opportunities and apply new technology rapidly. Similarly the cooperative and mutually supportive relations between the Japanese government and industry and the policies of the former appear to be more conducive to successful and timely innovation and technology application as the more confrontational, suspicious, and arms length relations between the government and industry in the U.S. for example.

For some time the Japanese were supposed to have derived much of their technological ideas from basic research performed in the United States and Europe. They were said to benefit from their ability to innovate and develop technology more rapidly because of the more cooperative and supportive environment, greater focus on commercialization, better feeling for and concern with customer or user requirements, and an ability to make more rapid decisions, including those requiring commitment of major resources for the development of technology.

Research or discovery of technological opportunity and innovation leading to technology application appear to be a much more continuous process in Japan than in the U.S. where R&D is often physically and organizationally segregated from engineering, design, and manufacturing, and thereby from technology development and innovation.

One issue of interest is the difference in the methods of acquisition of new technology and its commercialization. Technology acquisition involves functional, cultural, institutional, and organizational factors. As noted in the interesting study by R. S. Cutler[1], "the Japanese have a different language, a different thought process, and different social and business relations. To attempt to observe technology separately from its environment is to loose sight of this larger picture". In fact there is mounting evidence that the method of acquisition, innovation, and commercialization of technology in Japan is affected by strong cultural elements, as found by the authors of references [1] to [6]. The mechanisms involved in technology transfer, acquisition, and innovation for example uses multiple factors, involving functional behavior. In Japan there is a great feeling of trust between peers and between superiors and those who work for them.

Quality circles, peer review groups, and close personal cooperation without fear of competition are typical elements in Japanese research, development, and technology innovation organization which, by the way, usually come under one management unit which carries technology development all the way to commercialization. In addition government actively encourages formation of industry wide research and development groups (and often provides partial financial support for their work) for the purpose of developing new technology. These groups transfer the results of their research and innovation to the member companies for subsequent commercialization which may include additional development or innovation.

Prevailing anti-trust law and the competitive environment do not allow or encourage such cooperative technology development by U.S. firms. Recently the U.S. law has been changed to allow some research to be performed by industry consortia such as the Semi Conductor Research Corporation (SCR) and more recently a similar organization concerned with research in superconductivity.

Although in Japan the government acts as the technology development organizer and

---

[1] R. S. Cutler, "A Survey of High Technology Transfer Practices", ONR - Science Information Bulletin 13(2), 1988.

developer or cohesive industrial policy which individual firms use for their strategy determination, only 20% of all R&D funding is provided directly or indirectly by government, with about one-third of this spent at universities and government research institutions.

In Japan, as a result, there is a closely knit group of government, university, and industry experts who, through committee work and other associations, maintain close contact and exchange technological knowledge. As Cutler [1] points out, "Japanese companies do not suffer from the 'not invented here' syndrome, an attitude which stifles ideas from external sources", and is quite common in U.S. firms. Also the approach to commercialization of technology is quite different because it is less profit and more customer or market demand oriented. As a result, licensing of technology, even from or to competitors, is widespread in Japan.

In the area of technology communication, there seems to be much more open access to technology developments through committees, associations, and various cooperative arrangements for exchange of technological information. As a result, mistakes of one firm will only rarely be repeated by other firms. At professional meetings, knowledge of work or developments both inside and outside Japan will be exchanged. Some critics contend that all these are phenomena which emanate from the Japanese attitude about originality, which prefer to encourage a pattern of advances rather than completely new technology. Others maintain that this is largely the result of the fact that the Japanese term for learning (manabu) comes from the term for imitating (manebu) [1].

References

1. Cutler, R.S., "Impressions and Observations of Science and Technology in Japan", a report to the Japan-U.S. Educational Commission (Fulbright Foundation), University of Tokyo, Japan, May 1987.

2. Eagar, T. W., "Technology Transfer and Cooperative Research in Japan", ONR Far East Scientific Bulletin 10(3), 32-41, 1985.

3. Ouchi, W. G., Theory Z, Addison-Wesley Publishing Co., 1981.

4. Havelock, R. G. and Elder, V., "Technology transfer in Japan: An exploratory review", Center for Productive Use of Technology, George Mason University, June 1987.

5. Hill, C. T., "Japanese technical information: Opportunities to improve U.S. access", Congressional Research Service, October 1987.

6. Kobayashi, J., "Scientific creativity and engineering innovation in Japan", Symposium on Science in Japan, 153rd Annual Meeting, American Association for the Advancement of Science, Chicago, Illinois, February 16, 1987.

APPENDIX 11C - SUPPORT OF TECHNOLOGY DEVELOPMENT

Major changes have occurred in the support of technology developments during the last 30 years, a period during which mankind has made unprecedented technological advances. In 1960 the major World War II allies, the U.S., France, and the U.K., spent on technology developments in real terms and as a percentage of GNP significantly more than Japan and West Germany. In fact, U.S. expenditure for R&D alone (both public and private) exceeded that of the rest of the world in money terms. The U.S. spent 2.7-2.9% of GNP/year between 1960 and 1968. But this has since dropped by 2.1% of GNP between 1973 and 1978, and only grew slightly since. On the other hand Germany and Japan, both of whom spent a measly 1.2 and 1.35% of GNP during the early 1960s, have by now emerged as principal actors in R&D for technology development with both spending in excess of 2.7% of GNP by 1985. These percentages and the absolute expenditures for technology developments have since increased even further.

Today Japan (in 1988) spends 62% as much for R&D as the U.S., while Germany spends about 43% as much as the U.S. In other words, these two countries now outspend the U.S. These developments become even more striking if we consider only expenses for non-defense related R&D. In that case, the U.S. spent only 1.8% of GNP in 1985 compared to 2.8% of GNP for Japan and 2.65% of GNP for West Germany. Total non-defense expenditures for R&D, of these two countries was over 30% above that of the U.S. in 1985.

Another interesting issue is the number of scientists and engineers engaged in R&D, and the number of scientists and engineers graduated from universities per year. While the U.S. employed over 820,000 scientists and engineers in R&D in 1985, only 550,000 or about two-thirds worked on non-defense R&D. On the other hand, nearly all the 380,000 scientists and engineers engaged in R&D in Japan worked on non-defense R&D. Similarly, nearly all of West Germany's 120,000 scientists and engineers engaged in R&D worked on non-defense investigations. These two countries, as a result, employed nearly the same number in non-defense R&D as the U.S.

Yet, the number of engineers and scientists graduated by these two countries per year now exceeds that of the U.S. Similarly expenditures per R&D scientist or engineer in the U.S. were only $82,000 in constant 1982 dollars in 1985, versus $108,000 and $116,000 spent per scientist or engineer engaged in R&D in Japan and West Germany respectively.

The above-mentioned comparative support of technology development through R&D has changed even more since 1985. With the loss of about 70% of the U.S. dollar against the Japanese yen and nearly 60% lost against the West German mark between 1985 and 1989, the dollar equivalent of R&D expenditure is now skewed even more against the U.S. It is estimated that in 1988 the U.S. spent significantly less in 1988 dollar terms on non-defense R&D as Japan and West Germany, the major competitors of the U.S. in non-defense technology.

The organization of R&D and in particular cooperation among government, industry, and academic or other research institutions is quite different in the U.S. and Japan or West Germany. In Japan, government in cooperation with industry usually sets basic technology development policy and then convenes joint groups of industry and academic/research institution representatives who, under government guidance, develop plans for the development of new technology, including the R&D required. This approach has guided major technological advances in Japan, over many years and has resulted in rapid and coordinated technology developments. Once the basic R&D results are obtained, often through joint research, involving all interested industrial concerns and government or other research institutions, and the required inventions have been developed including confirmative innovation of the technology, the group is often dispersed with each industrial concern pursuing its own innovation, application, and often diffusion of the new technology.

The technology development path is similar though less structured in West Germany. Both approaches encourage industry-government cooperation in the establishment of technology policy and development. As a result, there is less overlap and wasted R&D, and speedier development. In particular, this approach assures greater integration of basic and applied research and development and a more continuous effort from R&D and discovery through the various stages of the innovation process. Continuity includes retention of experts involved in the R&D or discovery process to guide innovation.

In the U.S., an adversary relationship between government and industry largely curtails effective joint government, industry technology planning and development. There are under existing antitrust laws also legal constraints to intra industry collaboration in technology development.

Within industrial concerns, there is often a separation of R&D, and technology innovation and development. Most U.S. industrial concerns isolate their R&D from engineering, design, marketing, and manufacturing activities. This often causes discontinuities and inefficiencies in the development of new technologies in addition to major time delays in bringing the new technology to application or use.

There have been many studies and assertions on the role of military R&D in technology developments and the important role of technology spinoffs from defense research to non-defense innovation and ultimate use. While there are a few known spinoffs from technologies developed for defense, such as radar, there seem to be an equal number of spinoffs from non-defense research to defense innovation and use. A typical example is the use of laser and fiber optic technology. As a result, there are some claims that all R&D should be non-defense, as resulting technological breakthroughs will probably meet all legitimate technological needs of defense.

12. Environmental Impact and Future Role of Technological Change

Technological change affects not only the performance of products, processes, and services, but also their environment. There are many product technologies now which introduce completely new concepts as well as needs. In fact, many introduce or generate needs which people never thought they had or might have. Technology changes the behaviors of individuals, groups, and society. It introduces new challenges as well as new hazards. It unites people and drives them apart. It encourages confrontation of individuals, firms, and nations. The outcome of technological change is often perilous. It improves the means of communications, yet at the same time reduces the need for interpersonal and intersocietal communication, at least on a person-to-person level.

Modern technology is credited for our 'high standard of living' and at the same time is blamed for all the ills of modern society, its maladies, inequities, and disillusions. It is used to develop the most effective defenses, and the most destructive offensive weapons. Modern technology is responsible for most of the pollution of this globe's waters and air, the greenhouse effect, and the depletion of the ozone layer. In fact much of the methods of pollution imposed by human technology on this world since World War II, such as most of the radioactive and toxic pollution prevalent now, was unknown to mankind and the world before. We have introduced more pollutants into our environment since 1960 than in all previous human history. Yet new technology also provides effective means for cleanup and other environmental protection. Technological change is a mixed blessing and we must try to learn to live with it. We want its blessings, and these are not insignificant, but we often try to ignore its costs until later (and sometimes too late).

The effect of technology on culture is often irreversible as are impacts of technology on food and eating habits, food growing and distribution, transportation, and more.

The most difficult issue introduced by technological change is the valuing of its environmental impact and costs, not just by magnitude but also by rate and direction. There are usually short and long run effects and people may not, as a result, agree on the effect of a technological change, as some are more concerned with one than the other. A positive impact to some is often perceived as a negative environmental impact by others.

The environmental effects of technological change in developed countries accrued or accumulated over a two-hundred year period, or since the industrial revolution. Most industrialized countries adapted to these changes gradually over time, even though it changed their culture, style of living, and more. Technological change is much more difficult and destructive in the case of developing countries who often try (or are pushed) to introduce radical technological change in a short time, often a period of just a few years. The results have produced rapid economic advance in some cases, but have been disastrous in others because of the destruction of fibers of culture and tradition without compensatory economic improvements.

12.1 IMPACT OF TECHNOLOGICAL CHANGE ON DEVELOPING COUNTRIES

Developing countries (LDCs) usually obtain or adopt new technology long after its introduction into industrial countries. The reasons are not only that most new technology is developed in the industrial world which can afford to underwrite research and innovation, but also because most LDCs do not recognize the need for or potential application of new technology until it has been employed elsewhere for a long time. Even then, its introduction may be unneeded, useless, uneconomic, or counterincentive, and in many cases causes havoc with the local economy and social environment. Technology and development are supposed to be complementary, but recent history has shown them to be rather tenuous partners. Science and technology in LDCs often serve as status symbols more than contributors to economic and social development.

Academics and researchers in development of LDCs are often more preoccupied with scholarly treatise than the solution of actual problems by application of science and technology. This is understandable, as they compete for status and prestige with colleagues in developed countries who concentrate on scientific and technological fields with little application to the problems of technology and development in LDCs, such as: lack of basic economic research; affordable energy generation distribution and use; improvements in simple human skills and human resources in development; available health care; management training; effective distribution, and inefficient government and administrative structures.

Most new technology is developed for very narrow product, process, or service applications which offer unique and often highly specialized opportunities in developed nations. This technology often assumes availability of effective infrastructure, economic resource development, health care, educational systems, distribution of goods, and management ability.

For many years the concept of adoptive technology was used as a panacea, a way to develop technology specially designed for low technology countries - something simple and affordable. Unfortunately the adaptivity was usually conceived by well-meaning 'experts' with little, if any, experience in LDC's or knowledge of environmental and societal conditions. Examples abound and include introduction of simple reliable gasoline engine powered pumps to alleviate water shortages in subsaharan regions without an available local gasoline supply, to the supply of seaworthy North Sea coasters for interisland service on the calm seas among Indonesian archipelago islands with estuarial ports too shallow to be approached by such deep draft vessels. Adaptive technology was often grossly simplified technology which then lost much, if not all, its technological and operational advantage. It also sometimes meant transfer to LDCs of very mature, often obsolete, technology or technology that could not be used in industrial countries because of its operating requirements or environmental impacts. This often meant transfer of technology which was difficult to use in developed countries or that government outlawed or discouraged because it was environmentally unacceptable.

LDCs often tried to develop their own adaptive technology or transferred advanced technology to be adapted to their perceived needs. This also failed quite often, yet for different reasons. Nationalism and pride has driven some LDCs to purchase some of the most advanced technology often before it was fully developed and tried, only to have it abandon it because they lacked the financial and human capital needed for its development. Others tried to perform much of the development themselves (such as Brazil and India in computers, Indonesia and Brazil in small airplanes, and more) only to find this to be a rather expensive, and often counterincentive, approach.

The most successful adapters appear to be (NICs), countries who simply bought technology and knowhow often on a turnkey basis, even after achieving high degrees of technological proficiency themselves. Korean shipyards, for example, continued to purchase ship designs and many ship components, long after Korean shipbuilding had surpassed most shipbuilding countries in the world in quality, productivity, and quantity of shipbuilding output of all kinds of ships. They recognized that knowledge transfer which accompanies such technology transfer not only causes accelerated learning, but also permits concentration on technology adoption and improvements in its application instead of wasting valuable time and resources on reinvention and self-education which is counterproductive at the present time of rapid technological advance. Many countries who, for reasons of national pride, over confidence, or political dogma, choose to develop their 'own' technology in various fields found to their dismay that they usually fell further and further behind instead of catching up.

The most difficult issue facing technological change in developing countries is the effect on and by the environment and in particular the social environment. The major environmental factors are: (1) need for employment generation, usually among the unskilled and uneducated; (2) social structures which often hinder the education (and economic elevation) of lower class people; (3) cultural environments (and tradition) which discourage the educated from performing productive work; (4) political environments which emphasize 'prestige' investments and developments and not productivity enhancement; (5) traditional or religious constraints which prevent the productive use of certain human resources or types of employment (women, etc.); and, (6) barriers to foreign participation in the socio/economic policy.

Most developing countries have some or all of the above plus additional hangups. It is interesting to note that the newly industrialized countries (NICs) emphasized few, if any, of these environmental factors when they were at the development stage. They eliminated social taboos, encouraged education of all classes of people, attracted foreign investment and participation in their economic development, freely imported technology and knowhow, emphasized foreign values, and tried to relate them to their own cultural values, adjusted their political and social structure to facilitate modernization and efficiency, and moved towards income redistribution and other methods of motivation of the masses.

It is interesting to note the difference between the development of South Korea and the Philippines for example. Both had similar economies and populations in 1960. The GNP and per capita income of the Philippines at that time in fact surpassed that of South Korea. The Philippines also were, and are, much richer in minerals and usable land. They had an educational system and literacy surpassing that of Korea. They had received vast post-war reconstruction assistance and large foreign investments. True, they suffered under communist insurrection in the south, but then South Korea fought a war of survival in the fifties and is still effectively in a state of war with nearly 4 times as many men under arms than the Philippines.

Today South Korea's GNP (and per capita income) is a multiple of that of the Philippines. A similar development appears to take place now in China, a country just somewhat larger than India in population and area. The growth of GNP and per capita income of India surpassed that of China in 1980 soon after the state of liberalization of the Chinese economy, yet today China is rapidly overtaking India in economic terms, largely because of the increased economic freedom, reduced national constraints, emphasis on universal education, downplaying of traditional and political constraints to economic growth, and elimination of socio/political barriers. If the current trend continues, China's GNP and per capita income should be a multiple of that of India within 5 years.

In each of these successful cases technological change has managed to contribute most effectively to economic growth of the society by knowledge transfer and gradual adoption of the environment to the new technological opportunities. In other countries attempts are made to reinvent technology to suite the particular prevailing cultural environment. The Chinese, as people or individuals, did not change their attitudes or cultural values in the time between the revolution and today's new economic policy. But in the past they isolated themselves from the rest of the world and developed their own 'pure' technology. In some areas great progress was achieved, but in most fields the Chinese marched off on a tangent and found themselves many decades behind others when the doors of knowledge exchange were finally opened, with the advent of the new economic policy.

African countries present probably the worst situation. Although some African countries had an opportunity to advance economically and use technology effectively to this end, most failed dismally because they did not understand that technological change does not just consist of purchase of fancy, technologically advanced equipment. Technological change implies change in equipment, in manpower, management, education, social structure, government policy, and in all aspects involved in the change.

New technology requires a different approach to the use of manpower. Technological change cannot just be superimposed on or introduced into a socio/cultural environment, a traditional organizational and administrative structure, and sets of values and loyalties fostered by history. It requires adoption of the environment to the needs of the new technology and an effective choice of technology which should be allowed to assist in changing the existing environment. If the social and educational system cannot be changed over the short run, and if employment levels must be maintained, then advanced labor-saving process technology should not be introduced. Maybe process technology which improves the quality of the products and reduces costs without eliminating jobs, and which can be operated by the same unskilled workers should then be introduced as a first step.

This is not what was done in many African countries. New equipment and infrastructure, technology in advanced industrial countries was installed without a real attempt at environmental change. Most of these investments have been wasted. The dilapidated highways are retaken by the jungle, and broken down equipment lies everywhere for lack of maintenance or trained skilled operators.

Many valuable lessons can be learned from our development experience of the last 30 years. Technology can play an important role in economic development, but only if the environment is properly prepared for technological change, and the change is not just dumped on an existing non-technological change in economic development which requires broad-based change in the human resources, the administrative and social structure, training, and receptiveness of change. Technological change will interfere with tradition even if only raising the educational level of the young above that of their elders. The successful countries have been able to accept this change and admit that their new generation is smarter than their fathers. This can be done without a loss of respect for elders and tradition, if approached with courage. Unfortunately many people and societies dread and oppose social change, only to have it imposed anyway.

Traditional and other cultural values can live in harmony with technological change if they are accepted as complementary. If, on the other hand, tradition opposes technological change, both are known to suffer. This then is the issue of environmental impact which can make or break the success of technological change.

12.2 ENVIRONMENTAL IMPACT OF TECHNOLOGICAL CHANGE IN INDUSTRIAL COUNTRIES

We are all familiar with the impact of technological change on labor. In fact, labor unions have by and large resisted technological change because they assumed that it would reduce employment opportunities and require different types of labor than organized by their particular union. In addition, union leadership often associated technological change with changes in economic and therefore labor policy on a firm's or national basis. It does not matter than technological change in process, product, or service has in the long run always resulted in higher levels of employment, and often better average skills and income levels. In the short run, technological change does often cause employment dislocation, or even reduced job opportunities or unemployment. It may also result in social and other disruptions. Technological change in an industrial setting for example often impacts quite often on management requirements and the structure of the organization of the concerned enterprise. Technological change usually advances skill requirements, and moves decision levels upwards whereby the number of organizational levels in the management hierarchy are effectively reduced. The major environmental impacts of technological process change are:

1. change in work content changes and different skill level;
2. change in content and type of decisions required to manage process.

This would affect management structure.
3. change in the format and method of communications among peers and different levels of decisionmakers;
4. change in interpersonal relations and contact. Greater interpersonal exchange of experience.
5. changes in the role of individuals of the enterprise;
6. change in both the customers or users served by affecting their perceived demand; and,
7. change in the regulatory environment.

These impacts affect not only individual firms and organizations, but whole communities and society at large. Technological change in industrial nations has both a unifying and disruptive effect. it helps to assure better distributions of wealth, while at the same time increasing the poverty of the disadvantaged who are left out of the mainstream of employment. It increases the average skill level of industrial workers and the productivity of service employees, while downgrading the role of many service industry workers.

It helps improve the average quality of the physical environment while introducing new pollutants and other physically harmful substances or factors. It makes people economically more independent and socially dependent. In other words, technological change brings many advantages, but unless properly managed may cause undesirable effects or costs which in some cases may readily outweigh the benefits.

12.3 TECHNOLOGICAL CHANGE DECISION MAKING

Economic growth in recent years has been achieved largely through technology innovations in various fields of science and industry. From a firm's point of view, adapting new technology is one of the most important decisions managers have to make today. The adoption of the better and more efficient technology will make a firm better able to improve productivity to maintain its competitiveness and thus to sustain or increase its market share in the industry. The crucial questions to the managers are the timing, level of investment, method of technology acquisition, and the rate of technological change desirable to meeting the firm's objectives.

There are now decision analysis models available which will provide analytical support to the managers in making such decisions. Decision tree models can be used to evaluate technology options available for firm's adoption given various possible states (or important environmental factors) such as future demand and competitor's position. One shortfall of this type of model is that there may be some difficulty in estimating the gain due to management decisions and the dynamic changes in the competitor's position as they would also be aware of the opportunities for new technology adoptions.

Similarly, dynamic programming models can be effectively used in making decisions on the timing, level of investment, and the rate of technological change. Many of the inputs are treated as deterministic variables which may not represent the future realistically. This shortfall can be overcome to a certain degree by running several DP models using different values for each input. If stochastic DP models are desired, it may be difficult to include more than one stochastic variable.

Conditional probabilistic network or GERT models are designed to evaluate system performance resulting from the decisions on the technology adoption. The merit of using GERT models is their ability to consider conditional probabilities and the statistical distributions in resource usage and requirements. These models are especially useful in evaluating the methods and rate of technology acquisition (own research and development versus technology importing) and the interaction between production processes.

Usually, technological change is considered only in terms of its impact on productivity or costs. It would be interesting to examine the impact of technological change on other environmental factors and the organizational structure of the firm. For example, when the automation of a production process requires less manpower, but due to union rules or other commitments, the unneeded labor cannot always be fired. Thus, an efficient manpower planning and labor relocation planning should accompany the implementation of any planned automation process, if the benefits expected from such a technological change are to be attained.

Similarly, in measuring productivities, some guidelines are required to account for such qualitative factors as improved product quality and its impact on productivity.

12.3.1 *Future Management Challenges*. Technological decisions require technical expertise, the result of many years of training, as well as management ability. In other words, engineers or scientists may be required to deal with technological decisions. But in the U.S. few managers are engineers or scientists and few engineers have effective management skills. Therefore, technological change decisions are often delayed or delegated through a maze of experts or consultants from within and without the organization who 'advise' management. This is expensive, time consuming, and often less than decisive. In fact, such a process may result in built-in obsolescence or mismatch between new technology and actual requirements as the experts may be unfamiliar with the real requirements and the potential impact of technological changes advocated. In the U.S. more managers are technology adverse than in many other countries, particularly countries in the Orient, such as Japan or Korea. They perceive technology only as income-producing investments, and often plan technological change for short-time horizons to improve the balance sheet or other short-term financial objectives. Technological change, even if the new technology has a short expected life, must form part of a strategic management plan and not be a solution to a short-term or perceived problem, such as current loss of productivity or competitiveness.

The future role of technological change is that of an important strategic decision which considers and impacts on the function of a firm or organization, its endogenous resources and their use, and its environment, following its strategic objectives as guidelines. This requires a different type of manager - someone who appreciates and knows technology can effectively weigh its costs and benefits as a result of its expected performance under changing conditions, and who can weigh a technological change decision on the basis of the strategic objectives of the enterprise, its environmental constraints and impacts, and the effects on workers and other resources employed. To be an effective manager of technological change will, in the future, imply a successful manager, as no decision will be more important to the success of an enterprise than technological change decision.

These decisions will depend on both exogenous and endogenous factors, many of which may be outside the control of the manager. Among these are usually:

- demand for the product or process technology under consideration,
- cost of factor inputs and other resources required by the new technology,
- capabilities of the new technology,
- competitive pressures,
- endogenous and exogenous constraints,
- financial limitations, and
- technological risk.

The risk inherent in a change of technology can usually be estimated, but only as a consequence of impending developments, some of which may cause other problems.

12.4 TECHNOLOGICAL RISK MANAGEMENT

Technological change involves risk as does everything else, but the perception of that risk is amplified by greater uncertainty or lack of knowledge of what to expect, what may happen, and how things may develop. The risk of technological change involves many unknowns not only in the development and capability of the technology but also its performance, environmental, and societal effects, and more. The actual risks may be smaller than acceptance of the current technology, but because the risks are less known they will usually be perceived to be greater.

Therefore risk management must always be built into the design of a technological change in products, processes, or services. Risk management must involve unbiased and hopefully independent risk identification and risk assessment, including an effective causal tree which relates causes which introduce risks and the resulting or consequent effects. It involves a step by step procedure and mediation of conflicting risk assessments by experts. It must be value based to assure that unfounded or biased concerns not be given undue weight. Technological risk decisions involve

1. identification of risk factors or causes,
2. determination of events affected by these factors or causes,
3. risk assessment over the time and scope of the proposed technological change. This should be quantitative.
4. management rules or procedures to limit or reduce the identified risks,
5. development of estimates of costs of risk management (or reduction) and determination of impact on technology performance (i.e. does a 50% reduction in performance risk, result in a 10% degradation in performance), and
6. technological change decisionmaking.

Although risk management has been studied for over 20 years, very little is known as yet on how risk is understood and, more importantly, interpreted by individuals, and the public at large. In general, people appear to be willing to take greater risks if they believe they understand and can judge the risk involved. For example, the same people who habitually drive their car or use airplanes would not agree to live within 5 miles of a nuclear power plant, even though extensive statistics convincingly prove that the probability of a fatal or even injurious accident caused by driving or flying is probably 100 times as large as that possibly caused as a result of residing near a nuclear power plant. But most people are familiar with cars and believe to be expert enough at judging the risk. Furthermore, they often feel that a knowledge of the risk allows them to influence or even reduce it, though evidence shows that the average person will usually increase his risk in a known event by over confidence.

Public perception of risk has been studied by psychologists looking at cognitive processes. This includes group (including peer) interaction, when for example the experience of one group members affects the risk assessment of the group. Another factor which affects risk perception is the suddenness of the occurrence of the risk event. Risks which occur with warning are perceived to be more "acceptable" than those of equal magnitude, but which occur without warning.

In other words, risks are amplified by their occurrence among a peer group and by the suddenness of their occurrence. Another issue is the effect of communication and information on risk assessment. The more information is available on the cause and effect of the risk, including its statistics to closer public risk assessment will approach quantitative analytical risk assessment. The management of technological risk and particularly the risk of technological change therefore requires effective quantitative risk assessment and communication of the results as well as instruction in risk reduction.

Risk of technological failure is of increasing public as well as investor or user concern. The concern is usually not only with the economic cost of unavailability, inadequate or lack of performance, or unacceptable investment cost or development time, but also with the effect of failure of a new technology to perform effectively on the environment, public image, competitiveness, and market position of the innovator. Formal technology risk assessment should form an integral part of the planning and management of technological change. In general, we distinguish between two types of technological risk: 1. risk of technology performance in financial, operational, environmental, economic, and market terms; and 2. risk that technology will not perform, function, or solve problems for which it was designed.

Therefore risk in technological change is the potential for the realization of unwanted or unpredicted consequences, which in turn are caused by uncertainty in the various steps used in the evaluation of the need for new technology, R&D and innovation risk, investment (including procurement and foreign exchange) risk, technology performance, marketability, and price risk, and finally environmental and regulatory acceptance risk. Technology risk assessment usually consists of (1) technology risk determination which includes risk identification and risk estimation, and (2) technology risk evaluation which includes establishment of methods of risk aversion and estimation of risk acceptance.

Formal procedures exist for technology risk assessment which use stepwise cause and effect network analysis to determine the ultimate technology risk in qualitative or, if possible, quantitative terms. Technology risk assessment should usually be separated from technology risk evaluation and management is concerned. Risk assessment is often biased when performed by either technology proponents or potential "victims". Technology risk assessment is also complex because

1. There are many uncertainties in technology performance.

2. There are extreme value tradeoffs among proponents and opponents.

3. Neither proponents nor opponents are necessarily identified or identifiable.

4. There is often strategic posturing, political weighing involved. Furthermore secret agendas are often used which may result in the use of gaming approaches.

5. There is often well justified and unbiased disagreement among well-meaning experts about the uncertainties, performance, and value of a new technology.

6. Disputes and negotiations which often draw in secondary issues may be involved or become parts of the technology risk assessment.

Risk assessment should ideally be based on an effective statistically sound cause and effect or impact analysis which focuses on the impacts and consequences of events and the resulting costs and benefits. Risk assessment, as a result, should be descriptive and predictive, compared to risk evaluation which is normative and prescriptive. Technology risk assessment is performed to assure the general well being, while risk evaluation, which includes the results of risk assessment, is concerned with the development of a policy or strategy for technological change.

As noted, even experts often disagree in their opinions or predictions of risk in technological change. The reason is less real than cognitive bias, difference in the perception of the ability or role of the new technology, as well as various concepts of the way it could or should be used and its potential impacts in terms of benefits and costs. While some experts will consider only physical or measurable impacts, others may be more concerned with nonmeasurable or intangible, yet real, impacts.

Another issue is often the assessment of the impact of the rare risk event which, though extremely unlikely, might cause a catastrophe if and when it occurs. To deal with such issues, formal methods of risk assessment, risk evaluation, and risk management have been developed, as briefly presented in the next section.

Technology change risk assessment which deals with technology risk assessment, evaluation, and management consists largely of methods of planning for the unknowns to reduce uncertainty and thereby allow more effective management of risk of technological change. To this end we must identify:

1. potential problem areas, including their causes;
2. risks associated with these problem areas;
3. estimated levels and effects of risks;
4. possible alternative approaches to eliminate, mitigate, reduce, or otherwise control risks introduced by technological change; and,
5. plan of possible actions to that effect.

We usually have difficulty in dealing with uncertainty and unknowns, as it is conceptually disturbing to consider or plan for an unknown eventuality. In order to plan for technological risks, it is often desirable to divide the risks into

1. identified risks readily resolved or eliminated, given adequate effort, resources, and time are applied,

2. identified risks which need resolution where an effective problem solution may be available but has not yet been identified,

3. revealed risks, not properly identified, for which a method of resolution may have to be developed,

4. revealed risks for which there are little chances to identify (or develop) a method for risk resolution, and

5. residual unknown risks which must be expected to emerge, but cannot be planned for, nor can they be identified. Resolution approaches may, but will ordinarily not, exist.

Estimates of technological risk usually include expressions of uncertainty which will change as knowledge improves. On the other hand, technological risk is a natural component of technological change or as stated by Peter Drucker:

> "To try to eliminate risk in business enterprise is futile. Risk is inherent in the commitment of present resources to future expectations."

Technology risk assessment, the initial step in technological risk management, is usually defined as an analytical process capable of assessing the uncertainty and consequent risk involved in the introduction of the new technology, by providing an effective, traceable network from cause to effect of the risks involved, adequately to permit effective conceptualization and the development of possible mitigating procedures or problem solving measures.

In assessing risk it is convenient to organize our knowledge of the uncertainty relating to a technological change into two or more categories. Similarly the resolution or effort required to resolve the uncertainty is divided into several levels as shown.

	Known with Reasonable Certainty	Unknown and Uncertain		
		Presently Unknown	Completely	
Resolution Easily Achieved				Low Cost or Effort
Resolution Approach				Medium Cost or Effort
Resolution Unknown				High or Unachievable Cost or Effort

Next the risk resulting from the uncertainty is defined in terms of the probability and consequence of not achieving the performance of the new technology. The risk usually entails a risk of additional costs and/or loss of benefits in terms of revenues, market share, profits, etc. In other words, it is usually advisable to estimate the cost implications of the risk involved. Technological change includes risks of many types, some of which are quite difficult to quantify. There are risk factors introduced by the new technology:

1. Technology Performance Risk - The risk that the new technology will not perform as expected in operational, technological, and other terms.

2. Technology Timing Risk - The risk that the new technology is not available or available to perform as expected or needed if and when planned for use or introduced.

3. Technology Acceptance Risk - Risk that technology will not be accepted by user or market or introduce acceptance problems.

4. Financial Risk of New Technology - Risk that the new technology costs significantly more to introduce and/or use than expected.

5. Environmental Risk of New Technology - Risk that the new technology has a larger and/or different environmental impact than planned and acceptable.

6. Societal and Economic Risk of New Technology - Risk that new technology impact on training (skill), employment, work environment, societal relations, and local/regional economy in an unexpected manner which may introduce unplanned economic costs (or benefits).

Many of these risks are interdependent and dynamic. In other words, they affect each other and often change over time, with use of the new technology or as a result of endogenous or exogenous factors. It is usually convenient to consider and assess each risk factor separately and then introduce measures which represent dependency of interfacing factors. Similarly there are risk factors introduced by the close or broad environment in which the new technology is to perform. These include:

1. <u>Risk of Change in Requirements</u> - The risk that the requirements that the new technology is supposed to address or satisfy no longer exist or have changed if and when the new technology is introduced.

2. <u>Risk of Change in Market or Demand</u> - Risk that market or business direction will change.

3. <u>Risk of Change in Environmental Standards</u> - Risk that environmental standards will be different than expected when new technology is introduced.

4. <u>Risk of Change in Financial Capability</u> - Risk that financial capacity to develop, introduce, and use new technology will change by the time new technology is introduced.

In addition, there are various exogenous factors, such as work rules, political or trade developments, and more, which may impose risk to the introduction of a new technology. Finally, there are often significant secondary factors by which introduction of a new technology influences costs and benefits which cannot be ignored. Among these are the effect of the new technology on the organization, logistics, and role of an enterprise. The various risk factors are usually defined by the magnitude of their impact:

A. low, minor, moderate, significant, or high

and their attribute:

B. technical, cost, schedule, etc.

The attributes in turn are conveniently weighed to allow a total risk, or say a lack of achieved performance of a new technology, could be represented as follows:

		Technical	Investment Cost	Operating Cost
Magnitude	Weight	0.6	0.2	0.2
Low	1			
Minor	3		x	
Moderate	5	x		
Significant	7			x
High	9			
Total		2.0	0.6	1.4

<u>Total Performance Risk Weight therefore = 5.0</u>

Similar risk weights are computed for all other risks identified. To account for risk interdependence a factor which represents the modifying influence of one factor on another is introduced. These modifying factors are usually between 0.5 and 1.0.

Technology risks, as listed, can often be defined by their costs, benefits, or other quantifiable consequences with their associated probability distributions. To determine the total risk of introducing a new technology at a particular time, the costs of the various contributing risk factors are usually added and the linear bounds of the total uncertainty curves determined by assuming either total independence or dependence of the costs of the various risks. If c_1 and c_2 are random variables of costs then under independence

$$\sigma^2_{c_1+c_2} = \sigma^2_{c_1} + \sigma^2_{c_2}$$

and under dependence

$$\sigma^2_{c_1+c_2} = \sigma^2_{c_1} + \sigma^2_{c_2} + 2 \text{ cov } (c_1, c_2)$$

If random variables c_1 and c_2 are totally linearly correlated then

$$\sigma^2_{c_1+c_2} = (\sigma_{c_1} + \sigma_{c_2})^2$$

By mathematic induction these results can be extended to any number of random variables. As time progresses, the risk of a new technology changes. In part this is due to the decline of some and increase of other risks. As a result, the costs of the risk introduced by a new technology can usually be defined by a risk cost envelope as shown in Figure 12.1. As noted, the expected cost of risk or uncertainty increases and then decreases over time.

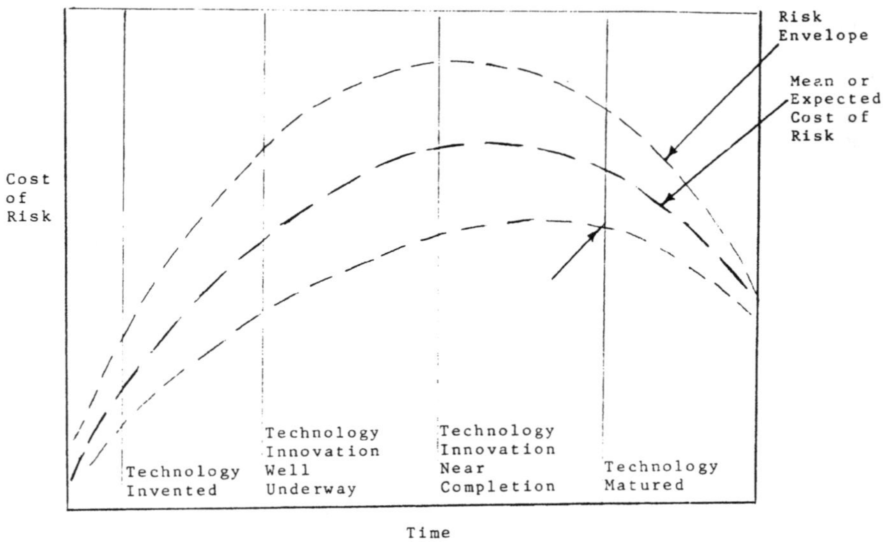

Figure 12.1. Technology Risk Cost Envelopes.

The envelope or deviation or spread of the cost of the technology risk, on the other hand, grows until innovation is underway, when it starts to decline rapidly, as most critical decisions are made during these early stages.

12.4.1 *Probabilistic Technology Risk Assessment.* The various risks introduced by new technology, and usually expressed in terms of the probability distribution of their cumulative costs, are conveniently identified by event or cause and effect trees in which effects or consequential events are expressed in terms of their costs. As mentioned before, the mean and the distribution of the cumulative effect cost curve changes over time for a technology and therefore with the timing of adoption or introduction of a new technology. Such cause and effect trees can be conveniently represented by conditional stochastic networks, such as GERT.

References

1. Canada, J.R., "Decision and Risk Analysis - a Review and Critique", Technical Papers, American Institute of Industrial Engineers, Twenty-fifth Anniversary Conference and Convention, Chicago, Ill., May 1973, pp. 37-40.

2. Hillier, F. S., "A Basic Approach to the Evaluation of Risky Interrelated Investments", in R. F. Byrne, A. Charnes, W. W. Cooper, O. A. Davis, and D. Gilford. Studies in Budgeting, North-Holland Publishing Company, Amsterdam, Holland, 1971, pp. 3-43.

3. Hillier, F. S., "The Evaluation of Risky Interrelated Investments", North-Holland Publishing Co., Amsterdam, Holland, 1969.

4. Wagle, B., "A Statistical Analysis of Risk in Investment Projects", Operational Research Quarterly, Vol. 18, No. 1, March 1967, pp. 13-33.

5. Renttinger, S., "Techniques for Project Appraisal Under Uncertainty", World Bank Occasional Papers No. 10, Washington, DC, 1970.

6. Frankel, Ernst, "Systems Reliability and Risk Analysis", 2nd Ed., Kluwer Academic Publishers, Dordrecht, The Netherlands, 1988.

7. Kenney, Ralph L., and Raiffa, Howard, "Decisions with Multiple Objectives" Preferences and Value Tradeoffs", John Wiley, New York, New York, 1976.

8. Quade, E. S., "Analysis for Public Decisions", Elsevier, New York, New York, 1975.

9. Quade, E. S. and Boucher, W. I., (eds.), "Systems Analysis and Policy Planning: Applications in Defense", American Elsevier, New York, New York, 1986.

10. Raiffa, Howard, "Decision Analysis: Introductory Lectures on Choices Under Uncertainty", Addison-Wesley, Reading, MA, 1968.

11. Sutherland, W., "Fundamentals of Cost Uncertainty Analysis", Report RAC-CR-4, Research Analysis Corporation, McLean, VA (AD881975), March 1971.

13. Strategic Management of Technological Change

Strategic management has become a subject of increasing importance. Medium- to long-term planning is essential now to position companies, governments, and other entities under conditions of continuous change not to develop rigid medium- to long-term guidelines, but to introduce a sense of direction based on overall medium- to long-term objectives. This is necessary to assure that opportunities are not missed.

Strategic management consists of the development and maintenance or updating of strategic plans which in turn serve to guide strategic decisionmaking. Long-term objectives of a firm are usually expressed in terms of market, market share, competitive position, role expansion, economic impact, etc., and not profit or other short-term goal. Technological changes have widely different potentials for contribution to a firm's strategic objectives.

They can affect competitive position, market, market share, market dynamics, profitability, employment, economic contribution, environmental effects, and more. New technology may change the position and role of a firm. It can change from a product to a process or marketing orientation or vice versa. New technology may make a weak, favorable, strong, or dominant impact on a firm's competitive position. Yet competitive position is increasingly driven by strategic technological change decisions. These will obviously differ with the type of firm, the major business objectives, and most importantly the stage of innovation or development of technology.

Strategic management of technological change uses as inputs (1) technology, condition, and firm condition audits, and (2) strategic objectives of the firm to derive a technological strategy which leads to technological investment strategies and technology development, introduction, and use programs. Strategic management is the development of plans and decisions which reflect and further the strategic objectives and general mission of the firm, government, or enterprise.

Strategic management must consider both internal and external factors. Management of technological change today is one of the most important issues involved in strategic management. Introduction or innovation of new technology can be a source of great opportunity or a major threat to a firm. It must therefore be managed to assure the consistency of the management of technological change with strategic objectives. Strategic management requires development of a set of strategic decision rules as shown in Figure 13.1.

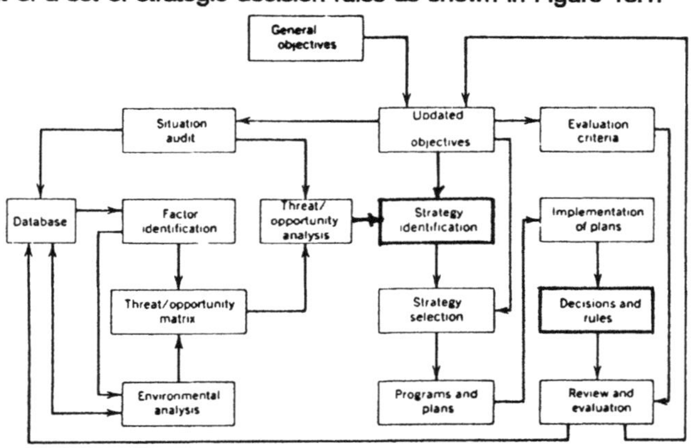

Figure 13.1. Process of Strategic Management.

These are derived from a strategy which in turn is developed from an analysis of the threats and opportunities existing or potential technological developments offer in light of the updated objectives of the firm. Another input into the threat opportunity analysis are the results of the firm's situation audit. Factors considered in establishing threats and opportunities of new technology often include:

1. cost and price of new technology;
2. technological advance;
3. competitive aspects;
4. obsolescence or maturity of new technology;
5. marketability;
6. state of development or innovation of new technology;
7. technological risk; and
8. others.

Threats and opportunities of technological change are usually expressed as an index of -1 to +10 or -100 to 100 and express the contribution (cost or benefit) of the technological change to the firm's strategic objectives. Threats and opportunities are conveniently represented by a matrix which is usually updated monthly or quarterly. It is common practice to divide strategic planning into a number of major stages such as (Figure 13.2):

1. environmental analysis and establishment of objectives, establishment of sensitivity of objectives;
2. resource and capability analysis, condition and performance audit;
3. problem identification;
4. policy development and threat/opportunity analysis;
5. development of strategic objectives, strategy and alternative strategy identification, and postulation of alternative strategies in terms of approach and impact;
6. results of forecasting, projection of potential results of alternative strategies;
7. selection of strategic projects and tactical analysis;
8. determination of effective and optimum tactics for implementation of selected strategy;
9. evaluation of selected projects and tactical approach; design of projects resulting from the implementation of chosen tactics;
10. project choice and plan development;
11. post audit, determination of results of tactical plans;
12. long-term corporate planning; and,
13. implementation of strategic plan.

A firm can use a reactive and proactive or innovative technology management strategy. It can act as a technology follower and rationalizer who only responds to market or competitive pressures, or it can become a technology leader or at least a developer of a technological niche.

Technological leaders usually develop new technology but often market the technology only as part of a long-term strategy and not as soon as the technology is available. They often leapfrog existing technologies or wait for strategic market opportunities to market technology developed by them.

The focus of a firm will depend on its

A. <u>Competitive Position</u> - weak, average, favorable, dominant

and

B. <u>Technological Position</u> - leader, niche developer, follower, rationalizer, acquisitor.

It will also be affected by the age or maturity of the firm (or industry). For effective strategic management of technological change, a firm must know its position, condition, and objectives as well as the state of technologies which may present threats or opportunities to it.

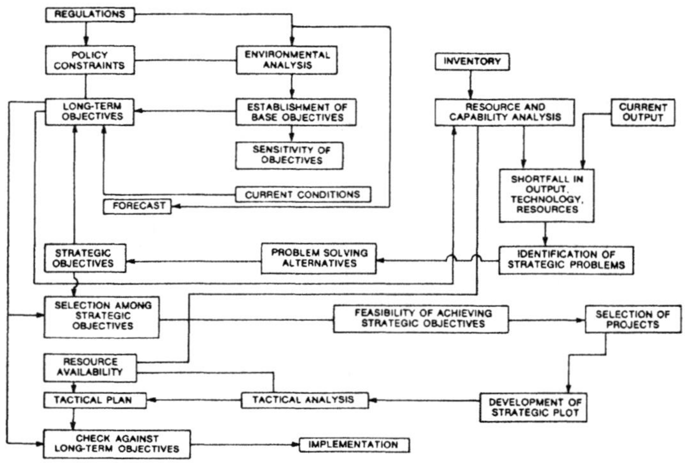

Figure 13.2. General Approach to Strategic Planning

INDEX

MIX
Papier aus verantwortungsvollen Quellen
Paper from responsible sources
FSC® C105338

If you have any concerns about our products,
you can contact us on
ProductSafety@springernature.com

In case Publisher is established outside the EU,
the EU authorized representative is:
Springer Nature Customer Service Center GmbH
Europaplatz 3, 69115 Heidelberg, Germany

Printed by Libri Plureos GmbH
in Hamburg, Germany